AF251740

Boron and Gadolinium Neutron Capture Therapy for Cancer Treatment

Boron and Gadolinium Neutron Capture Therapy for Cancer Treatment

Narayan S Hosmane
Northern Illinois University, USA

John A Maguire
Southern Methodist University, USA

Yinghuai Zhu
Institute of Chemical and Engineering Sciences (ICES), Singapore

Masao Takagaki
Kyoto University, Japan

World Scientific

NEW JERSEY · LONDON · SINGAPORE · BEIJING · SHANGHAI · HONG KONG · TAIPEI · CHENNAI

Published by

World Scientific Publishing Co. Pte. Ltd.

5 Toh Tuck Link, Singapore 596224

USA office: 27 Warren Street, Suite 401-402, Hackensack, NJ 07601

UK office: 57 Shelton Street, Covent Garden, London WC2H 9HE

British Library Cataloguing-in-Publication Data
A catalogue record for this book is available from the British Library.

**BORON AND GADOLINIUM NEUTRON CAPTURE THERAPY
FOR CANCER TREATMENT**

ISBN-13 978-981-4338-67 7
ISBN-10 981-4338-67-2

Typeset by Stallion Press
Email: enquiries@stallionpress.com

Printed by FuIsland Offset Printing (S) Pte Ltd Singapore

This book is dedicated to **Professor M. Frederick Hawthorne**, Director of the International Institute of Nano and Molecular Medicine at the University of Missouri-Columbia and to all of the Cancer Care Professionals around the globe. Professor Hawthorne, who is one of the founders of polyhedral boron chemistry, has dedicated his later career to curing cancer through the innovative, state-of-the-art therapy, known as Boron Neutron Capture Therapy (BNCT) which is the subject of this book. Much of the progress in this experimental endeavor rests on Professor Hawthorne's work. All of the **Cancer Care Professionals**, such as physicians, nurses, nurse-assistants, clinicians, interns, scientists, councilors, psychiatrics, and educators alike, have dedicated their lifelong career in caring for the cancer patients. Their dedication to our global society is hereby recognized and saluted.

Foreword

M. Frederick Hawthorne

It is an honor to be asked to introduce this concise and stimulating book which describes the current state of both ^{10}B and ^{157}Gd neutron capture as applied to medicine. Since I have invested much of my professional career in the development of the now well-known field of polyhedral borane chemistry and its interface with boron neutron capture therapy (BNCT), I am particularly pleased that this book forcefully describes the chemistry of neutron capture target species. This subject has always been given weaker emphasis than other technologies comprising BNCT which has contributed to the current diminution of progress.

The fact that boron chemistry was virtually unknown when BNCT was initiated in the 1950s probably initiated today's stalemate. Once identified as a useful post-WWII venture which demonstrated a peaceful use of atomic energy, the BNCT concept was rapidly applied to glioblastoma multiforme therapy (GBM). A clinical trial was initiated using existing neutron assets. GBM is a dangerous disease found in an easily accessible location, the head. BNCT target compounds of a useful nature would not be available until polyhedral borane dianions, carboranes, metallacarboranes, and 3-center chemical bond theory were established for creating a new supporting field of chemistry (1960 onward). At that time, international clinicians had reached far ahead of needed chemistry, pharmacology, and biology with clinical trials of GBM using poor neutron sources, inferior boron target compounds, and dosimetry/treatment planning which was just emerging. No tumor types other than GBM were extensively investigated and the fate of BNCT was rapidly fixed by the monetary and professional investments made in what proved to be unsuccessful GBM therapy. Now, 50 years later, one finds the same three boron target compounds allowed for human use, no target compounds created after 1958 that were ever used in ^{10}B-enriched form with tumor-bearing animals and little interest in BNCT in the medical community since BNCT and GBM are usually tightly associated in context. Current BNCT research has essentially slowed to a near standstill since the public apparently believes that GBM therapy is the only useful fate for BNCT and funding agencies are not supportive of a resurgence of a "failed" concept. The fact is that BNCT was never adequately explored while tied to GBM therapy. What is the problem? Simply put, it remains the identification of acceptable target compounds for tumors other than GBM although many such compounds are known which 20 years ago gave excellent results in small animal biodistribution experiments but never examined beyond that point. These species remain as candidates for further development when pitted against tumor types such as head or neck, prostate, breast, and brain metastases. Even arthritis may be treatable using BNCT-mediated synovectomy.

Gadolinium neutron capture therapy is noted for the very high neutron cross-section of the ^{157}Gd isotope (240,000 barns) which greatly enhances the probability of ^{157}Gd/thermal neutron interaction. Such interaction is less destructive than ^{10}B neutron capture since the therapeutic debris of a nuclear event is principally Auger electrons having a lower lethality than the ^{10}B fission products, ^{7}Li^{3+} and ^{4}He^{2+}. Consequently, the most effective use of a ^{157}Gd target places the metal in a chelate which is then placed near the DNA in the cancer cell nucleus to achieve double-strand DNA rupture. This process operates under the shadow of Gd heavy metal toxicity which must be avoided as in the case of MRI contrast agents. Due to these complexities not encountered in BNCT, GdNCT has made less progress. If one accepts the fact that BNCT has not yet been evaluated with the most effective types of boron agents targeted against tumor types other than GBM, the BNCT future again appears to be promising. It is my hope that the novel and effective boron agents described in this volume, as well as others, are subjected to thorough evaluation including the irradiation of tumor-bearing animals. This work must be successful in order to arouse financial support of a new agent synthesis and evaluation program either government or privately financed. Perhaps the best hope for improved agents and their delivery to tumor lies in the evaluation of the most effective of older agents, the application of nano particulate delivery, and more imaginative clinical procedures. One could summarize by noting that many pieces of the BNCT puzzle are lying around, but a rigorous and vigorous evaluation mechanism is needed in order to proceed. It is my hope that this book will stimulate such a recovery of the BNCT field.

Preface

Cancer is one of the most lethal diseases known. The death rate from this disease was over eight million in 2008 (~14% of the total deaths worldwide) and is predicted to rise to about 12 million (~18% of the total deaths worldwide) by 2030. This book covers an emerging type of cancer therapy, called neutron capture therapy (NCT). It is both a history of the development of NCT, including its successes and failures, as well as a scientific discussion of its principles. Unfortunately, until quite recently, the failures greatly outnumbered the successes. The basic idea of the therapy has been known for almost as long as the neutron has been known. It revolves around preferentially localizing compounds containing ^{10}B or ^{157}Gd in cancer cells and irradiating with low energy neutrons. These two nuclei have a high propensity to absorb a neutron that initiates a cytotoxic event. Conceptually the idea is intriguingly simple. This was perhaps its greatest weakness; the idea seemed too simple. It was pursued before we had a good understanding of the intricacies of cellular dynamics and structure. The method itself is brute force — localize and blast away. The subtleness comes in how do you preferentially insert these compounds in cancer cells and supply the correct amount of neutron flux, while avoiding healthy cells. The answers are derived from a number of different fields: chemistry, cell biology, computer science, medicine, and nuclear physics. This therapy is a group effort; this is noted by the diversity of the scientific journals listed in the references, everything from standard chemical, biological and physics journals to the more specialized radiation oncology journals. Since the contributing areas are diverse, one area could be limited by the others and *vice versa*. This book is an attempt to bring together a number of different fields that share a common goal: to cure cancer.

In reviewing the contributors, certain names will appear more often than others, to these we owe special thanks; without their contributions, this field would not have advanced. Chief among these is Professor M. Frederick Hawthorne, Director of the International Institute of Nano and Molecular Medicine at the University of Missouri in Columbia, Missouri. Fred has made significant contributions to both boron hydride chemistry and Boron Neutron Capture Therapy (BNCT). His pioneering research has laid the foundation of much of the current work in carborane cluster compounds and their applications to BNCT. For his many past, present and future, contributions to this field, we gladly dedicate this book.

In the preparation of this manuscript, several individuals have been unusually helpful, especially Ms. Kate Krise, Mr. Hiren Patel and Dr. Amartya Chakrabarti. Chapter 2 is taken almost directly from Kate's research paper entitled "What is Cancer?" It has been modified to fit into the format of the book, although we tried to maintain as much as possible

Kate Krise

Hiren Patel

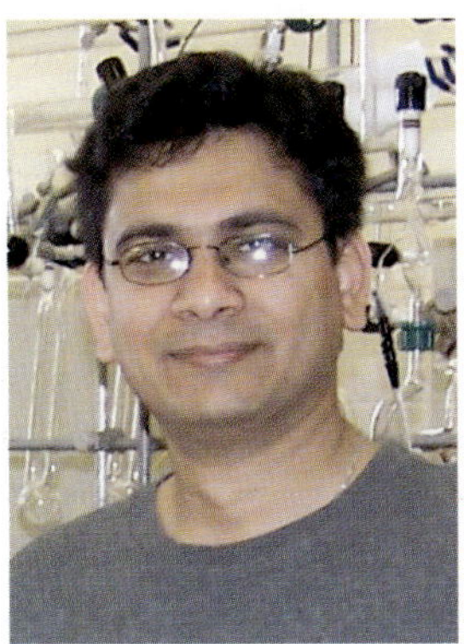

Amartya Chakrabarti

the effectiveness of Kate's original writing. Hiren Patel, who is an artist of exceptional caliber, as well as one of our research students, volunteered in drawing almost all of the figures, schemes, and tables in the book. Dr. Amartya Chakrabarti oversaw the work of these two young researchers; we express our sincere thanks to Amartya for this help. Our special thanks go to Professor Dennis N. Kevill and Mrs. Sumathy Hosmane of Northern Illinois University who kindly agreed to read the manuscript and made invaluable suggestions.

Last but not least, we wish to express our thanks to Sandhya Venkatesh, Senior Editor, World Scientific Publishing. If it were not for her persuasive ability to speak the same native Indian language, Kannada, as does Dr. Hosmane, he would not have committed to this venture, and, in turn, would not have browbeaten the other three authors into joining him in this venture.

About the Authors

Narayan S. Hosmane, PhD

Narayan S. Hosmane was born in Gokarna near Goa, Karnataka state, Southern India, and is a BS and MS graduate of Karnataka University, India. He obtained a PhD degree in Inorganic Chemistry in 1974 from the University of Edinburgh, Scotland, under the supervision of Professor Evelyn Ebsworth. After a brief postdoctoral research training in Professor Frank Glockling's laboratory at the Queen's University of Belfast, he joined the Lambeg Research Institute in Northern Ireland, and then moved to the USA to study carboranes and metallacarboranes. After postdoctoral work with Russell Grimes at the University of Virginia, in 1979 he joined the faculty at the Virginia Polytechnic Institute and State University. In 1982, he joined the faculty at the Southern Methodist University, where he became Professor of Chemistry in 1989. In 1998, he moved to Northern Illinois University and is currently a Distinguished Research Professor and Inaugural Board of Trustees Professor. Dr Hosmane is widely acknowledged to have an international reputation as "one of the world leaders in an interesting, important, and very active area of boron chemistry that is related to Cancer Research" and as "one of the most influential boron chemists practicing today." Hosmane has received numerous international awards that include but are not limited to the Alexander von Humboldt Foundation's Senior US Scientist Award twice; the BUSA Award for Distinguished Achievements in Boron Science; the Pandit Jawaharlal Nehru Distinguished Chair of Chemistry at the University of Hyderabad, India; and the Gauss Professorship of the Göttingen Academy of Sciences in Germany. He has published over 270 papers in leading scientific journals.

John A. Maguire, PhD

John Maguire was born in Alabama, received his BS Degree in Chemistry from Birmingham Southern College and his PhD Degree in Physical Chemistry in 1963 from Northwestern University under the direction of Ralph G. Pearson. He joined the faculty of Southern Methodist University as an Assistant Professor of Chemistry and is now Professor of Chemistry.

From 1979–1981, he was Dean of University College and General Education at SMU. His research interests are in the structure and properties of intercalation compounds and in the bonding, structure and reactivity in carborane cage compounds. He has received several awards including outstanding professor award for teaching in 1975 and in 1995, and Perrine Prize for research. He has published over 140 scientific peer-reviewed papers. In 2002, he was appointed University Distinguished Teaching Professor.

Yinghuai Zhu, PhD

Yinghuai Zhu, born in China in 1969, received his PhD in 1997 from Nankai University in Organometallic Chemistry. He continued his studies as a postdoctoral fellow in boron chemistry at Northern Illinois University and Southern Methodist University in 2000. He then joined the faculty at the Institute of Chemical and Engineering Sciences, Singapore, in 2002 and is currently a Research Scientist. His research focuses on the design and synthesis of organometallic complex-based catalysts for classical organic reaction and biomass transformation. Other research interest includes constructing functional material for biological application.

Masao Takagaki, MD, PhD

Masao Takagaki is a neurosurgeon in Japan who received her MD and PhD from Kyoto University. Her main research field is in the area of neutron capture therapy (NCT) for the treatment of cancer. She has studied NCT for malignant brain tumors at the Research Reactor Institute of Kyoto University (KURRI) as an Assistant Professor from 1990 to 2001. She now continues her research into NCT as Professor of Nursing at Aino Gakuin College at KURRI.

Contents

Chapter 1

Introduction

Malignancy is a life-threatening condition in which tumor cells are of a highly invasive character and penetrate normal organs and exhibit an obstinate resistance against cancer therapy. Glioblastoma multiforme (GBM) in the brain, undifferentiated carcinoma of the thyroid gland, small cell lung cancer, and malignant melanoma are typically malignant and their five-year survival rates are very low, 5–10%. It is an earnest wish to control them in human being. However, malignant tumors are highly invasive, they easily metastasize to the lymph nodes and other organs at a very early stage. As a rule, a malignant brain tumor (e.g. GBM) does not invade extracranially, but it invades throughout the brain at an early stage. Generally, malignancies are treated by multidisciplinary modalities, such as extirpation + radiation + chemotherapy as shown in Fig. 1.1. As for radiation therapy, malignant tumors are resistant to photon therapies, such as X-ray and γ-ray, thus high linear energy transfer (LET) radiation such as heavy ionizing particles and neutron capture therapy must be employed.

Neutron capture therapy (NCT) is a binary cancer treatment in which a cytotoxic event is triggered when an atom, placed in a cancer cell, absorbs a low energy, thermal, neutron. It is a seemingly ideal form of therapy, in that it kills cancer cells while sparing healthy cells.

Although many of the references cited in this book are quite recent, the basic concept of neutron capture therapy was first proposed in 1936 by Gordon L. Locher[1] when he formulated his binary concept of treating

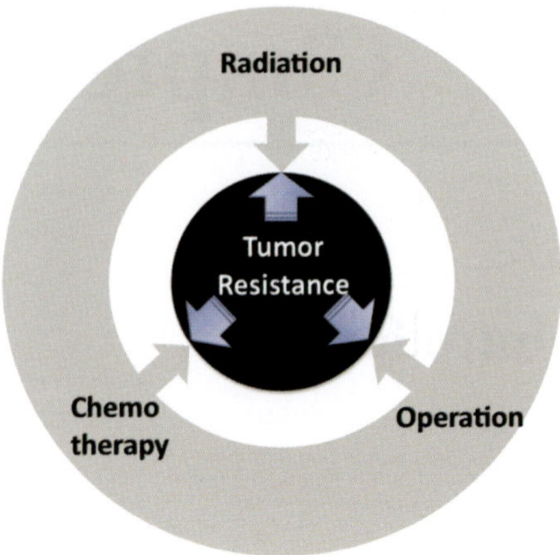

Figure 1.1   The treatment strategy for malignancies.

cancer: enough knowledge of the remarkable behavior of neutrons has been accumulated through physical research to enable the prediction of certain biological effects, and to see, in a general way at least, certain therapeutic potentialities of this new kind of corpuscular radiation. In particular, there exist the possibilities of introducing small quantities of strong neutron absorbers into the regions where it is desired to liberate ionization energy (a simple illustration would be the injection of a soluble, nontoxic compound of boron, lithium, gadolinium, or gold into a superficial cancer, followed by bombardment with slow neutrons.

This remarkable approach to cancer treatment was proposed shortly after the discovery of the neutron. It not only outlines the general approach but also gives the two elements, boron and gadolinium, which are currently studied as NCT agents. These two elements are unusual in that several of their isotopes have very high abilities to absorb thermal neutrons. The probability that a particular isotope will absorb a thermal neutron is given by its nuclear capture cross section, σ_{th}, measured in barns (b = 10^{-24} cm^2). This should not be taken as a measure of physical size, but rather as a measure of the probability of a nucleus absorbing a neutron. In general the capture cross section of a nucleus increases as the velocity of the bombarding neutron decreases; slow neutrons spend

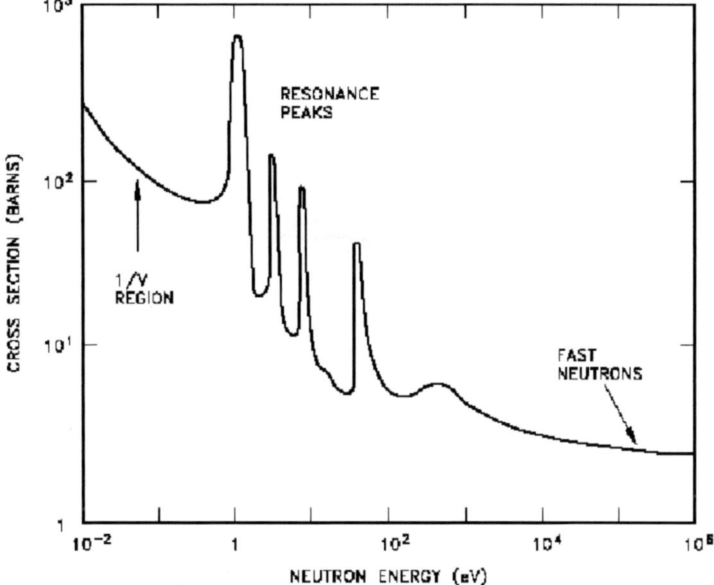

Figure 1.2 Change in the neutron capture cross section with the energy of the neutrons.

more time around a nucleus and hence are more likely to be captured. Figure 1.2 shows the change in cross section as a function of neutron speed.

In the low-energy region (thermal neutrons) the cross section varies as 1/v (v = speed of the neutron), this is the "1/v region" and the cross section is large. This is followed by a region of resonance peaks where the cross sections rise and fall sharply, depending on whether the energy of the neutron closely matches the discrete quantum levels of the nucleus. The fast neutron region, where the cross sections are very small, follows this. In NCT, one is interested in the region of thermal neutrons.

Boron-10, has an unusually high capture cross section of 3838 b, while the isotopes ^{155}Gd and ^{157}Gd have even larger ones of 60,900 b and 255,000 b, respectively, for thermal neutrons. These cross sections are orders of magnitude higher than those of the elements that constitute the

Table 1.1   Thermal neutron capture cross sections of common biological tissue.

Nuclide	Thermal Neutron Capture Cross Section sth[b] 1 barn = $10-24$ cm^2	Weight % in Mammalian Tissue	Nuclide	Thermal Neutron Capture Cross Section sth[b] 1 barn = $10-24$ cm^2	Weight % in Mammalian Tissue
^{1}H	0.333	10	^{31}P	0.18	1.16
^{12}C	0.0035	18	^{32}S	0.53	0.2
^{14}N	1.83	3	^{35}Cl	32.68	0.16
^{16}O	0.00019	65	^{39}K	2.1	0.2
^{23}Na	0.43	0.11	^{40}Ca	0.4	2.01
^{24}Mg	0.0053	0.04	^{56}Fe	2.57	0.01

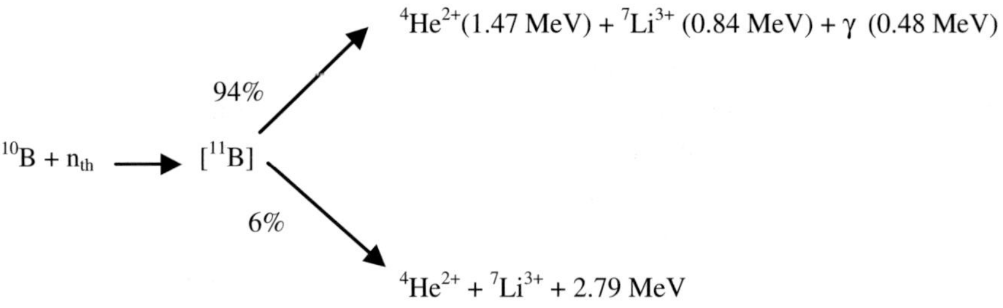

Scheme 1.1   Boron neutron capture reaction.

bulk of biological tissue (see Table 1.1.) The magnitude of these "cross sections" depends on nuclear structure, for example, σ_{th} for ^{11}B is only 5.5×10^{-3} b. This is significant in that natural abundant boron is only about 20% ^{10}B.

On absorption of a thermal neutron (E < 0.4 eV) by ^{10}B, an excited ^{11}B is formed that almost immediately ($\sim 10^{-12}$s) undergoes a fission reaction producing two high-energy heavy ions, ^{4}He^{2+}(α-particle) and ^{7}Li^{3+}, and a low energy γ-ray (see Scheme 1.1).

The high kinetic energy ^{4}He^{2+} and the ^{7}Li^{3+} transfer their energies over a very short distance, ~ 4–9 µm in biological tissue, this LET is of the order of a cell diameter. Therefore, if sufficient compounds containing

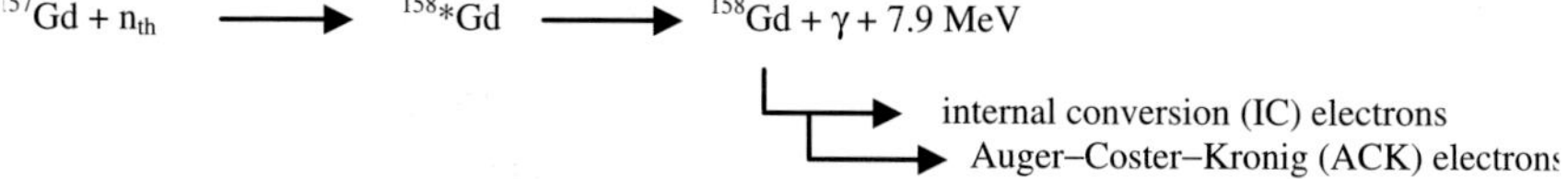

Scheme 1.2   Gadolinium neutron capture reaction.

^{10}B can be preferentially absorbed in a cancer cell, and bombarded with thermal neutrons, cell damage would be confined to that particular cell, sparing neighboring healthy cells. This is the basis of all BNCT treatments.

The gadolinium neutron capture (GdNC) reaction, $^{157}Gd(n,\gamma)^{158}{*}Gd$, is more complex than that of boron, and it is not a fission reaction. Of its seven stable isotopes, gadolinium has two isotopes that are of interest to NCT, ^{155}Gd (σ_{th} = 55,000 b) and ^{157}Gd (σ_{th} = 255,000 b). The ^{157}Gd has the highest thermal neutron capture cross section of all the stable isotopes in the periodic table. The GdNC reaction initiates complex inner-shell transitions, as shown in Scheme 1.2, that generate prompt γ emission displacing an inner core electron which, in turn, induces internal-conversion (IC) electron emission and finally Auger–Coster–Kronig (ACK) electron emission along with soft X-ray and photon emissions.[3–8] The low-LET γ-rays have an average energy of 2.2 MeV with a range of several centimeters. The IC electrons possess average energies of around 45 eV, paired with a range of several millimeters in physiological tissue. Finally, the range of the very low energy ACK electrons is only several nanometers in aqueous solutions. Although low in absolute energy, the ultra-short range of the ACK electron energy deposition effectively makes the ACK radiation component of the GdNC reaction a high-LET-type and is radiologically the most relevant component of the GdNC.[9.] The ACK electrons provoke high LET-type damage with a mean free path of 12.5 nm. Since this ionizing radiation is limited to molecular dimensions, it is essential to place the gadolinium atoms within the DNA helix to induce significant DNA damage in cancer cells.

Both clinical and chemical studies related to BNCT have progressed since the early 1950s. Boron compounds are easily incorporated in organic

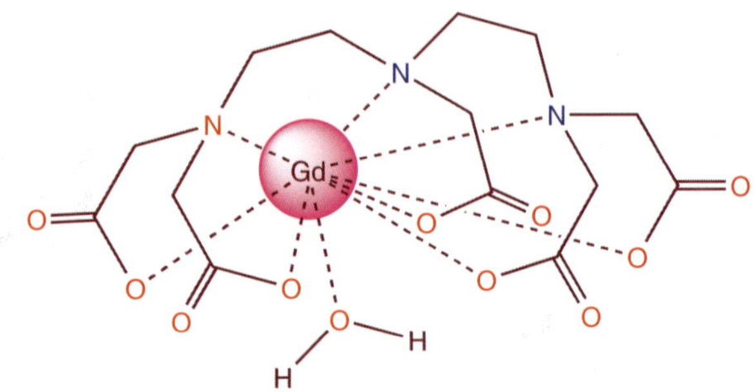

Figure 1.3   MRI contrast agent $[Gd(DTPA)(H_2O)]^{2-}$ magnevist.

structures and their chemistry has been extensively investigated.[10] On the other hand, the toxicity of Gd^{3+} had prevented its use until the development of suitable complexes of Gd^{3+}, such as gadopentetate dimeglumine, $[Gd(DTPA)(H_2O)]^{2-}$ (see Fig. 1.3), and their use as magnetic resonance imaging (MRI) contrast agents in 1988. As a consequence, there is much more literature available on BNCT than there is on GdNCT.[11]

Chapter

What Is Cancer?

2.1. Historical Background

Everyone is linked to cancer in some manner, whether it is through their own personal experiences and battles, having a family history of the disease, or having observed a friend struggle through the process. The word "cancer" is a general term that is used in reference to over 100 different diseases. In 2008, there were about 8 million deaths worldwide attributed to cancer and it is estimated that the number will rise to approximately 12 million by 2030. Normal cells in the body grow, divide, and die. Cancerous cells, however, result when cells begin to grow out of control; they become "immortal." Normal cells become cancerous due to mutations caused by damage/change in DNA. This happens more often than one would realize; it is estimated that in human cells the average number of DNA damages per person is about 800/hr. The DNA repair enzymes remove most DNA damages, but unfortunately these enzymes are not 100% effective. Some genes control critical cell functions such as cell growth or programmed cell death (apoptosis); these are called proto-oncogenes. When these are altered by mutation, they can give rise to oncogenes that can lead to cancer. In the case of cancer, the damaged DNA is not properly repaired; the cell is allowed to generate with its abnormalities existing. The abnormal cells then continue to replicate, passing its damaged DNA code onto its daughter cells.

While the damaged DNA that can result in cancer is sometimes inherited, in many cases, the resulting abnormal cells are due to the environment and lifestyle choices. For instance, cigarette smoking increases the risk of abnormal cell growth in the lungs. Cancerous tissue can form in any region of the body. In most cases, a solid tumor (mass of cells) is formed. The out-of-control growth of cancerous cells poses many threats. Due to their rapid growth and immortality, the cells can quickly spread (metastasize) to other locations in the body, causing further differentiation and damage.

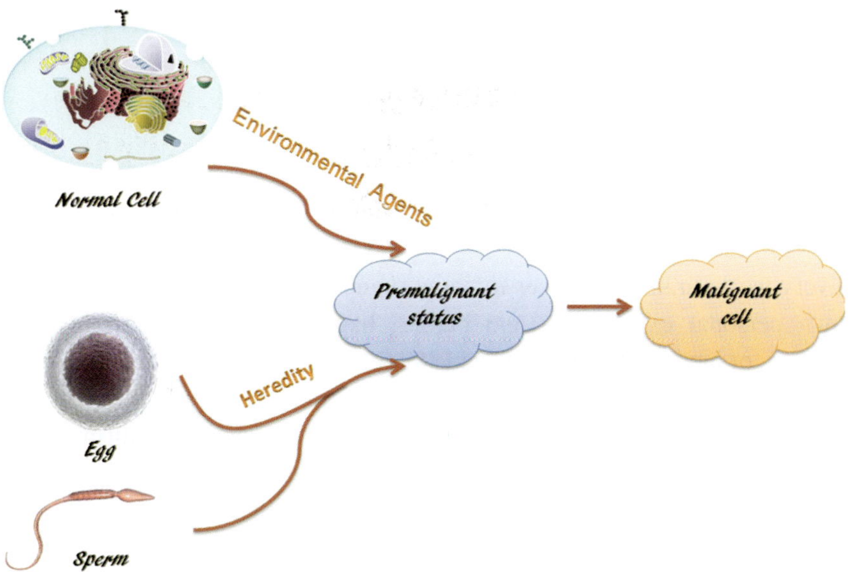

Figure 2.1   Origin of cell mutation.

Not all cancers exist as a solid tumor; there are other forms that can be referred to as "liquid tumors."[12] A common type of liquid tumor is leukemia (a cancer of the blood, which is usually characterized by an increase in white blood cells). Cancers of the blood and blood-forming organs, such as bone marrow, are classified under "liquid tumors."

2.2. Characteristics of Tumors

A tumor can be classified as "benign" or "malignant" (see Scheme 2.1).[13] A simple way to remember which type of tumor is more severe is to think that benign sounds like "be nice." A malignant tumor is the type of tumor that is capable of spreading to different areas of the body, whereas a benign tumor will not spread to other regions. Benign tumors are non-metastatic (unable to spread to different areas) because they are contained within a fibrous capsule that prevents them from invading surrounding tissue. Benign cells are differentiated, meaning that they closely resemble the normal tissue from which they are derived. Undifferentiated or "anaplastic" cells vary in size and shape and are usually piled on top of one another in an unorganized manner and are not contained within a capsule. Anaplastic cells are of the type that can penetrate blood vessels and lymphatic tissue, spreading to different regions of the body. The resulting secondary tumor is referred to as a "metastasis."

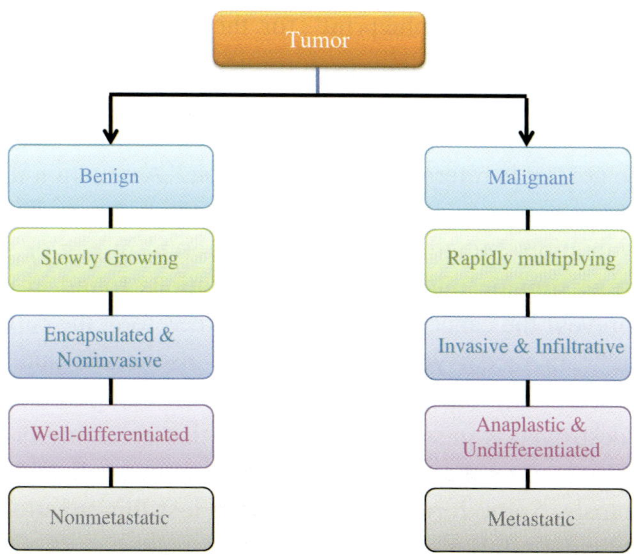

Scheme 2.1   Classification of tumors.

2.3. Origin of Cancer

Research is still being conducted that is aimed at trying to understand how different types of cancer are formed and how to effectively treat them.

What is understood is that cancer results from damaged DNA. DNA carries genetic codes that are unique to each person. While it serves many roles, it has one primary function. The DNA contains the codes for every protein in the body. The code acts as a blueprint for the production of new proteins. Each code, a separate and distinct gene, is a specific sequence of nucleotides. A nucleotide is composed of a sugar (ribose or deoxyribose), phosphate and a base (adenine, guanine, cytosine, thymine, or uracil).

Malignant tumors can result when mitosis or the genetic code is disrupted, resulting in damaged DNA. This can be a result of environmental conditions such as the presence of toxins, tobacco smoke, or viruses, which can cause a chemical change in the components of the gene. The change is often due to a change in the order of the bases or a change in a base, a deletion or insertion of certain base pairs, etc. When the gene is altered, the "blueprint" for the protein will be changed, often resulting in the protein performing a function not originally intended.

When the change in DNA can be inherited from one generation of cells to the next, it is referred to as a "mutation." A mutation that can be passed from a parent to a child through the egg or sperm is called an inherited mutation. Mutations that are not present in egg or sperm are referred to as somatic mutations. Some mutations found in cancerous cells can prevent the cancer cell from dying; therefore, the cell becomes "immortal."[14] In recent years, research has discovered that, in many types of cancers, the "blueprint" for directing the aging or damaged tissue to be destroyed is missing. Normal cells that are aging or damaged should be directed to undergo a type of programmed cell death referred to as apoptosis. Some forms of cancer have lost this function and are able to live indefinitely.[13]

2.4. Environmental Agents

Agents in the environment, such as tobacco smoke, viruses, radiation, drugs, and chemicals can cause DNA damage that will result in the production of cancerous tissue. A cancer causing agent is referred to as a "carcinogen."

2.4.1. *Radiation*

Radiation is a type of energy that is in the form of a wave.[15] It can come from many sources, such as the sun, an X-ray, or through a radioactive substance. Just like any other carcinogen, it can cause damage or mutations to DNA. For instance, survivors of the atomic bomb explosion in Hiroshima and Nagasaki exhibited a much higher than normal incidence of leukemia and other forms of cancer. Leukemia is a common form of radiation-induced cancer that results from bone marrow's sensitivity to radiation. Ultraviolet radiation emitted from the sun has also been linked to various forms of skin cancer. This is particularly dangerous for those who have lightly pigmented skin.

It is important to note that not all forms of radiation are hazardous. When used properly, radiation therapy can actually be utilized to treat cancer. Radiation therapy involves dosages many thousands times higher than the dosage used for a diagnostic X-ray. Therefore, there are risks associated with the radiation treatment, such as secondary cancers forming as a result of the therapy. When undergoing radiation therapy, the risks and benefits of the treatment are weighed before treatment is administered. In most cases, the benefits greatly outweigh the risk.

An important distinction that separates the risk associated with radiation is whether it is ionizing or nonionizing. Ionizing radiation has a higher frequency and is able to remove electrons from an atom/molecule or directly damage DNA. Forms of ionizing radiation would include gamma rays, X-rays, and ultraviolet radiation from the sun. The type of radiation that would be emitted from a cell phone, television, or microwave is of a lower frequency. It is nonionizing and it has not been linked to causing cancer. However, the World Health Organization (WHO) has recently raised questions as to the long-term effect of heavy cell phone use.

Many forms of radiation can be found naturally in the environment, such as cosmic rays from the sun and stars and radioactive elements in the soil and water. A person who spends larger periods of time outdoors will most likely be at a higher risk of developing a form of skin cancer. Those who live in a high altitude area, such as Denver, Colorado, would have increased exposure and risk. The largest source or naturally occurring radiation for most people is radon (^{222}Rn). It is a colorless, odorless gas that forms as a result of radioactive decay in the ^{238}U series. Radon is an alpha particle emitter ($t_{1/2}$ = 3.8 days) that is frequently found in the basements of commercial buildings and homes. It is the second leading cause of lung cancer, following cigarette smoking, and it is estimated that radon causes between 15,000 and 22,000 deaths per year in the United States. When purchasing a home, it is often a protocol to have the basement checked by the US Environmental Protection Agency (EPA) for the presence of radon.

2.4.2. *Chemical carcinogens*

Chemical carcinogens can be found in the atmosphere, medications, foods, water supply, and even drugs. Recently, the EPA has stated that approximately one-third of the active ingredients used in pesticides are toxic and one-fourth are carcinogenic.[16] This means that we are potentially exposing ourselves to carcinogens through the food we eat. Tobacco smokers can experience a "triple threat," firstly due to the pesticides used to grow tobacco, secondly due to the low levels of radiation which can be found in the soil due to the fertilizers, and thirdly due to the carcinogenic effects to the lungs due to smoking. Even prescription drugs can lead to cancer in the organs being targeted. For example, drugs containing estrogen can cause cancer by stimulating proliferation of the cells that line the uterus.

2.4.3. *Oncogenic viruses*

Some viruses can be carcinogenic (Table 2.1).[17] In 2002, the WHO International Agency for Research on Cancer estimated that 20% of human cancers were caused by infection.[18] A few viruses have been directly linked to specific types of cancer, such as the human T-cell leukemia virus (HTLV) which presents as adult leukemia, the human papillomavirus (HPV), which can cause cervical, head, and neck cancers, and the hepatitis B virus (HBV), which causes liver cancer. The discovery of which particular viruses are linked to specific cancers and diseases has resulted in research and production of preventative vaccinations. For instance, the first vaccine that was produced to prevent a form of liver cancer (hepatocellular carcinoma) was the hepatitis B vaccine.[19] Another vaccine (Gardasil®)[20] has recently been manufactured and distributed to

Table 2.1 Oncogenic DNA viruses.

Virus	Symptoms and Disease	Cancer
Human papillomavirus	Warts, including STD genital warts	Cervical cancer
Hepatitis B	Liver inflammation, jaundice, liver cirrhosis	Hepatocellular carcinoma (liver cancer)
Hepatitis C	Chronic infection, scarring of the liver, esophageal varices, gastric varices	Hepatocellular carcinoma (liver cancer)
Epstein–Barr	Mononucleosis (glandular fever), lymphoproliferative disorders	Hodgkin's disease, Burkitt's lymphoma, nasopharyngeal carcinoma
Human T-lymphotropic virus	Demyelinating diseases, tropical spastic paraparesis	T-cell leukemia, T-cell lymphoma
Kaposi's sarcoma	Red/purple/brown/black nodules on the skin, mouth, gastrointestinal tract, and respiratory tract	Cancer of lymphatic endothelium
Merkel cell polyomavirus	Skin lesions	Merkel cell carcinoma (skin cancer)

young women to aid in protection from certain strains of HPV and, therefore, help prevent the occurrence of cervical cancer.

Tumor producing viruses are often referred to as "oncogenic viruses," and fall into three categories: RNA viruses (such as hepatitis C virus), DNA viruses (such as HPV), and retroviruses (such as the human T-lymphotrophic virus and hepatitis B virus), which can have both DNA and RNA genomes.[13] Some viruses become tumorigenic by first infecting a cell and then persist as a circular episome or plasmid that replicates separately from the host cells DNA (such as the Epstein–Barr virus and Kaposi's sarcoma). Other viruses become tumorigenic only after they become integrated into the host cell's genome (such as polyomaviruses and papillomaviruses).

2.4.4. *Heredity*

Not all cancers develop due to environmental factors; some people are genetically predisposed due to a parent passing defective DNA through their gametes (egg or sperm) onto their children.[13] Some examples of known inherited cancers are polyposis coli syndrome (which causes polyps to grow in the colon and rectum) and retinoblastoma (which causes a tumor on the retina of the eye). Retinoblastoma is a childhood cancer that results due to a mutation in the tumor suppressor gene Rb. Detection of hereditary cancers can usually be performed by a simple genetic analysis obtained from cells, such as blood cells.

In many cases, research leads to the belief that these types of tumors arise due to an inherited or acquired abnormality in a group of genes called suppressor genes. A normal individual's suppressor genes would help promote a cell in becoming more specialized; it would regulate cellular growth and suppress oncogenes from causing cancer. When the function of a suppressor gene is lost, cell division becomes out of control and malignancy can result. An example of a suppressor gene is the p53 gene. A mutation in the p53 gene can result in a higher risk of childhood sarcoma, leukemia, and brain (central nervous system) cancers.

If there is a prevalence of certain forms of cancer, genetic screening is often recommended for other family members. Since inherited changes can be detected in all tissues in the body, a sample can easily be obtained for analysis.

Approximately half of the forms of cancer that result in death originate in the lung, breast, and colon. Although there are over 100 different individual forms of cancer, they can be classified into one of three broad groups. The largest group is carcinomas, followed by sarcomas, and then mixed-tissue tumors.[13]

Carcinomas are derived from epithelial tissue that lines the inner and outer surfaces of the body. Almost 90% of malignancies are classified as carcinomas.[21] The types of tissues that carcinomas are derived from include the skin, glands, digestive tract, urinary, and reproductive organs. The most common types of carcinomas are breast, lung, colon, and prostate. In 2008, they accounted for ~38% of deaths due to malignant neoplasms, worldwide.

Sarcomas are the second most common group of cancer. They are derived from supportive/connective tissues such as bone, fat, and cartilage. Sarcomas include cancers such as leukemia, because blood-forming tissues are considered as a connective tissue. The least common groups of tumors are the mixed-tissue tumors. They are derived from tissues that are capable of differentiating into both epithelial and connective tissue (see Table 2.2).

Table 2.2 Groups of cancers and origins of tissues.

Groups of Cancers and Tissues from which They are Derived
Carcinomas
Gastrointestinal Tract, Glandular Tissue, Kidney and Bladder, Lung, Reproductive Organs, and Skin
Sarcomas
Bone, Muscle, Cartilage, Fat, Fibrous Tissue, Blood Vessel Tissue, and Nerve Tissue
Mixed-Tissue
Kidney, Ovaries, and Testes

2.5. Pathology

The process of properly identifying whether a worrisome lump is cancerous involves several steps. There are many aspects to be considered, such as whether the lump is malignant or benign, how abnormal or differentiated the cells are, what type of organ/tissue the abnormality is located in, and if it has infiltrated other regions of the body. Once all of those aspects have been properly identified, a course of treatment can be determined. The process of accurately identifying all of the above is crucial for determining the correct treatment.

In order to ensure the proper etiology and treatment, the skills of a pathologist are usually enlisted. A pathologist is a consultant physician/scientist who studies disease.[22] They use laboratory analysis to aid in diagnosis and treatment. Pathologists are able to observe changes in organs, tissues, and cells, evaluate the extent of the disease, and predict the course the disease will most likely take. Their knowledge is invaluable in determining the proper course of treatment. Pathologists will often attend a surgery, in order to obtain tissue samples and analyze them during the surgery. Their findings can give the surgeon insight as to whether or not a tissue is necessary to repair or remove.

After a pathologist has analyzed a specimen, they compile their findings in a pathology report, which can be easily read by other physicians and medical staff. The pathology report identifies what type of cancer was found, the grade of the cancer, tumor size, whether the cancer is invasive, and how far it has spread. The report might also include other factors that could give insight into the aggressiveness of the cancer.

A pathologist first looks at a tumor with the naked eye; this is referred to as a "gross examination."[23] When analyzing macroscopically, appearance and characteristics such as size, weight, and texture are noted. The pathologist then would cut the tissue into thin slices in order to microscopically examine the tissue with a conventional light microscope. The tissue is then measured. The distance from the edge of the mass to the edge of the specimen is measured; this is referred to as a "margin."[24] A positive margin indicates that cancerous cells continue toward the edge of the specimen. A negative margin indicates that the cells do not extend to the edge of the tissue. Typically additional surgery is necessary with a positive or close margin.

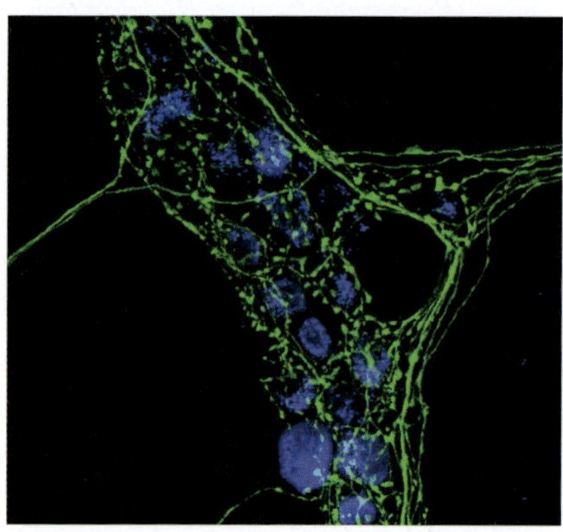 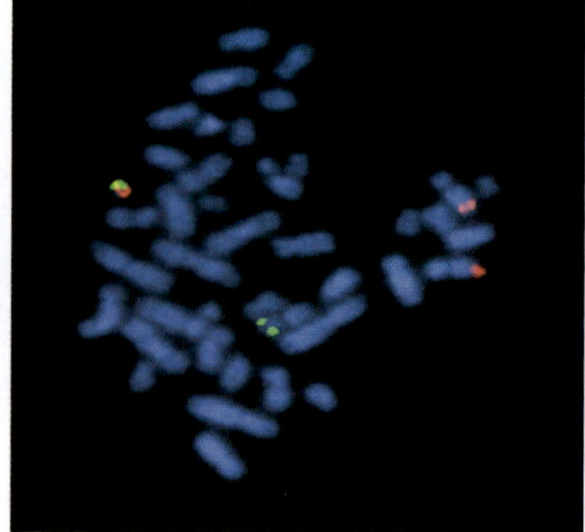

(a) Immunohistochemistry labels individual proteins, such as TH (green) in the axons of sympathetic autonomic neurons.

(b) A metaphase cell positive for the bcr/abl rearrangement (associated with chronic myelogenous leukemia) using FISH. The chromosomes can be seen in blue. The chromosome that is labeled with green and red spots (upper left) is the one where the wrong rearrangement is present.

Figure 2.2   (a) Immunohistochemistry (IHC) and (b) Fluorescent *in situ* hybridization (FISH).

There are various sensitivity techniques used to analyze a tissue. A common test involves immunohistochemistry (IHC).[25] IHC exploits the process of an antibodies ability to bind to an antigen (e.g. proteins). For example, many tumors depend on hormones. IHC is able to recognize the presence of hormone receptors, in order to determine if a tumor could be treated with antihormonal therapy. For instance, one of the first antihormonal therapies used to treat breast cancer was an antiestrogen called tamoxifen.[26]

Another sensitivity technique often used is fluorescent *in situ* hybridization (FISH).[27] FISH is used to detect the presence or absence of specific DNA sequences on chromosomes by using fluorescent molecules to color genes. It is used to determine how many copies of a specific gene are present in a cell. FISH in combination with IHC are very useful to reach a diagnosis.

2.6. Grading

When a cancer is graded, the aggressiveness and how abnormal the tissue appears microscopically are analyzed.[28] Most cancers are graded based on how normal/differentiated or abnormal/dedifferentiated the cells are. The features of interest are the size and shape of the nucleus of the cell, the fraction of cells dividing, and the arrangement of the cells. Grading is performed in a laboratory setting after a tissue biopsy is obtained. The most common grading system is the Gleason System. The number range on the Gleason System is from 2 to 10, with 2 being the most normal looking cells. Grades of less than 4 usually means that the cells look similar to normal cells and that the cancer is likely to be nonaggressive, grades 4–7 are intermediate stages in which the cells do not look normal but the cancer is of a slower growing, less aggressive nature, grades 8–10 indicate aggressive, rapid growing cancers. Other grading systems are used that vary in their numerical ranges; however, it can be assumed, when reading any grading system, that a lower number indicates a less severe form of cancer.

2.7. Staging

There are two forms of staging. There is clinical staging, which estimates the extent of cancer based on the imaging tests (X-rays, CT scans,

MRIs, ultrasounds, and PET scans), biopsies, and physical examinations. The second form of staging is pathologic staging, which is usually obtained during surgery. The two forms of staging in combination are used in order to better understand the severity of a cancer. The common elements that are considered when classifying the stage of a cancer include:

- the location of the primary tumor;
- the tumor size;
- the number of tumors;
- if the cancer has spread to the lymph nodes;
- how closely the cancer cells resemble normal tissue; and
- if the cancer has infiltrated other regions of the body.

Stage	Description
Tumor (T)	
TX	Primary tumor cannot be assessed
T0	No evidence of primary tumor
Tis	Carcinoma *in situ*
T1	Tumor invades subepithelial connective tissue
T2	Tumor invades corpus spongiosum or cavemosum
T3	Tumor invades urethra or prostate
T4	Tumor invades other adjacent structures
Node (N)	
NX	Regional lymph nodes cannot be assessed
N0	No regional lymph node metastasis
N1	Metastasis in a single , superficial, inguinal lymph node
N2	Metastasis multiple or bilateral superficial inguinal lymph nodes
N3	Metastasis in deep inguinal or pelvis lymph node(s) unilateral or bilateral
Metastatis (M)	
MX	Distant metastasis cannot be assessed
MO	No evidence of distant metastasis
M1	Distant metastasis cannot be assessed

Figure 2.3   Criteria for cancer staging (*AJCC Cancer Staging Manual*, 5th edn. (1997)).

The most commonly used staging system is the TNM (Tumor, lymph Nodes, Metastasis) staging,[29] which was developed by the American Joint Committee on Cancer (AJCC). It is mostly based on three different characteristics of the cancer. The T refers to the size of the tumor and location in which it originated. The number following the T, normally given as T0 through T4, indicates the size or level of invasiveness into neighboring structures. The higher the number, the larger the tumor or the more it has grown into nearby tissues. The N refers to whether the lymph nodes draining the area of the primary tumor are involved and the extent of involvement. N0 means that there is no lymph node involvement, while N4 indicates extensive lymph node involvement. M is used to indicate the presence or absence of metastasis. Three values following the M indicate different things. MX would signify that metastasis cannot be evaluated. M0 means that no distant metastases can be found and M1 means that there are distant metastases.

The overall stage grouping uses Roman numerals, I–IV, along with 0.

Stage 0: Carcinoma is *in situ*, that is, they have not started to invade neighboring tissue.

Stage I: Cancers are localized in one part of the body.

Stages II and III: Cancers are locally advanced.

Stage IV: Cancers are metastasized or spread to other organs.

Depending on the different types of cancers, the criteria for staging differ. For example, in colon cancer, T3N0M0 is a stage II type of cancer, whereas, bladder cancer T3N0M0 would be stage III. Some forms of cancer do not use the TNM staging system, for instance, brain cancers do not because of their tendencies to spread to other regions of the brain instead of lymph nodes and other parts of the body. Leukemias also are not TNM-staged because the blood and bone marrow throughout the body are affected.

While TNM is the most common staging system, other staging systems exist. For instance, the International Federation of Gynecologists and Obstetricians (FIGO)[30] have developed a system (which closely resembles TNM) that classifies cancers of the female reproductive organs. Other systems also exist for cancers of specific organ systems.

A commonly misunderstood part of staging is that the stage of a cancer does not change over time, regardless of whether a cancer progresses. A cancer is always referred by the stage in which the cancer was originally discovered. The degree of metastasis is just added to the original stage. A cancer will often undergo "restaging," which involves performing additional tests to assess the cancer after treatment.[31] Restaging is done to measure whether or not a cancer has responded to a treatment in order to determine if a cancer has returned. In most cases, when restaging is done, a new stage is not assigned. If a new stage is assigned, a letter "r" will be placed before the previous stage to indicate the change.

2.8. Surgery

In many cases, the best chance to cure cancer, if it has not spread to other areas, has been surgery.[13] Surgery can be used to repair, remove, diagnose, treat, and relieve symptoms. It can also be supplemented by other treatments, such as hormone therapy, chemotherapy, and biological therapy. Removal of all or part of a tissue plays a crucial role in the primary diagnosis of determining if a tumor is malignant or benign and to establish the degree of severity. Some genetic conditions such as familial polyposis, once identified, often require preventative surgery to remove the colon or rectum due to the high risk of developing colon cancer. Sometimes surgery cannot remove the entire tumor, but it can still be helpful when used in combination with other treatments such as chemotherapy and radiation. When only a portion of a tumor is removed, to improve quality of life, it is referred to as "debulking." Debulking can aid in relieving pressure on a nerve or bone or to unblock an obstruction.

2.8.1. *Various forms of cancer surgery*

2.8.1.1. *Laparoscopic surgery*

It is a minimally invasive operation that is used to reduce recovery time, hospital stay, and the risk of complications.[32] It involves a fiberoptic scope and surgical instruments that are inserted through small incisions, in order to perform the necessary surgery. Most laparoscopic surgeries involve insufflating (inflating) the abdomen with carbon dioxide gas. Carbon dioxide is used because it can be absorbed by the body or removed by the respiratory system and is not flammable.

2.8.1.2. *Robotic surgery*

It is a newer form of surgery that a few hospitals have begun to introduce. It has proven to be incredibly precise and allows smaller incisions to be made.[33] It is a less invasive procedure that requires less recovery time for the patient. The da Vinci Surgical System®, which is manufactured by Intuitive Surgical, has proven to be very advanced surgical system. A surgeon is able to sit at a control panel and operate the da Vinci Surgical System by means of two foot pedals and two hand controllers. It is able to translate these movements into more precise micromovements thus improving accuracy and decreasing tissue damage. While da Vinci robotics proves very promising, it will most likely take a long period of time for most hospitals to adopt this system, due to the $1.3 million cost of the equipment and the training required.

2.8.1.3. *Radiofrequency ablation (RFA)*

Radiofrequency ablation (RFA) is a form of therapy in which a needle-like RFA probe is placed inside a tumor, and radiofrequency waves are then passed through the probe, heating the tumor and destroying it.[34] It is usually an outpatient procedure used to treat tumors located in the kidney, lung, and bone. Since it does not directly stimulate nerves or heart muscle, it can often be used without the use of general anesthesia. RFA can be used independently or in combination with chemotherapy and radiation therapy to increase effectiveness.

2.9. Radiation Therapy

As previously mentioned in the causes of cancer, radiation is a contributing factor that can lead to development of cancer but it can also be used to treat cancer.[13] Radiation therapy works by delivering a maximum amount of ionizing radiation to a tumor, while attempting to minimize the exposure to normal/healthy tissue. Newer techniques utilize high-energy beams that improve the focus of the beam, therefore limiting the damage to healthy tissues, although some DNA damage cannot be entirely avoided. It actually is necessary to include some margin of normal tissue in order to ensure that the entire tumor is irradiated. A possible unfortunate side effect of radiation therapy is that the DNA damage to normal tissue can lead to the development of secondary cancers.

2.9.1. *Types of radiation therapy*

2.9.1.1. *Brachytherapy*

It is a form of radiation that involves the source of radiation (seeds or wires) being placed inside or next to the tumor.[35] It has proven to be a very effective treatment for cervical, prostate, breast, and skin cancers. Brachytherapy affects only a small region surrounding the source of radiation. It can be placed directly inside the target tissue (interstitial brachytherapy) or near the target tissue (contact brachytherapy). Sometimes the placement can be permanent (such as in prostate cancer treatment) or temporary.

2.9.1.2. *Conformal radiotherapy*

It is also referred to as 3D conformal radiotherapy which allows doctors to see their treatment in three dimensions.[36] By using a computer program, multiple radiation beams are able to conform more closely to the shape of the tumor. This allows for a higher dose of radiation to be given more accurately while reducing possible side effects. The dimensions of the tumor can be planned entirely by means of a CT scan or with the aid of an MRI or PET scan. There is a specific form of conformal radiotherapy called intensity modulated radiotherapy (IMRT) that enables the radiation beam to not only be conformed around the tumor but also the dose to be altered, depending on the shape. A combination of these two features allows for the central portion of the tumor to receive the higher dose or radiation.

2.9.1.3. *Stereotactic radiosurgery*

It uses a high amount radiation to treat brain and neck tumors. The Gamma Knife™ is one form of radiosurgery in which a specialized helmet is surgically fastened to the head, to prevent movement, in order to accommodate the proper delivery of radiation.[37] With Gamma Knife radiosurgery, the source of radiation is kept stationary while the patient is rotated. Due to the necessity of keeping the area of the tumor stabilized, Gamma Knife is mostly limited to the head and neck. Another form of radiosurgery is the Cyberknife, which works in a similar fashion as the Gamma Knife except the Cyberknife does not require an immobilizer.[38] The radiation is directly delivered from the outside of the body into the tumor.

2.9.1.4. *Proton beam therapy*

It is a newer form of radiation therapy in which the positively charged proton is used to treat cancer.[39] The general mechanism through which it works is that the proton pulls negatively charged electrons out of the orbitals of molecules surrounding cancerous cells. The act of removing the electrons is called "ionization." The process of ionization causes damage to the DNA and therefore cellular destruction. An added benefit of proton therapy is that protons can be accelerated to a specific velocity that is determined by the physician in order to deposit the maximum energy at the desired location. As the protons moved through the tissues, they slow down near the tumor, thus causing ionization of the orbiting electrons in the cancerous cells. The resulting cellular damage is more difficult for the cancerous cells to repair in comparison to normal cells. Therefore, an even greater percentage of the destruction of cancerous cells results.

2.9.1.5. *Neutron capture therapy*

It is a bimodal therapy in which a compound containing ^{10}B or ^{157}Gd is preferentially localized in cancer cells and irradiated with thermal neutrons, thereby initiating a cytotoxic sequence. It is the subject of this book and will be covered in detail in the following chapters.

2.10. Chemotherapy

Chemotherapy (sometimes referred to as just simply "chemo") is a type of systemic therapy that involves administering drugs intravenously.[40] It is often the standard treatment for patients with choriocarcinoma, testicular cancer, acute lymphocytic leukemia, breast cancer, and Hodgkin disease. Chemo is often given following surgery as a supplemental treatment to ensure full eradication.

A negative aspect to chemotherapy is that the rapidly dividing cancer cells are not the only ones affected. The other cells in the body will also suffer. Some normal cells such as bone marrow, digestive tract, hair follicles, nails, and the mouth are affected. However, the benefit of destroying the cancer outweighs the other normal tissue damage that results as side effects.

2.10.1. *Common chemotherapeutic agents*

2.10.1.1. *Alkylating agents*

Alkylating agents damage DNA and prevent it from dividing and replicating. There are three methods by which an alkylating agent can prevent DNA synthesis.[41] They can attach alkyl groups, form cross-bridges, or introduce mispaired nucleotides. Any of those methods ultimately results in the inability for replication to continue.

2.10.1.2. *Antimetabolites*

These are structurally similar to components of the building blocks of proteins and DNA structures.[42] They are able to be mistakenly taken up by the body instead of a metabolite, thus causing the resulting cell to be unable to carry out its vital functions. The downside to using antimetabolites for treatment is that they are only effective in specific portions of the cell cycle, therefore they can only be used for certain treatments. Antimetabolites are often used to treat cancers of the ovaries, blood, breast, lung, and GI tract.

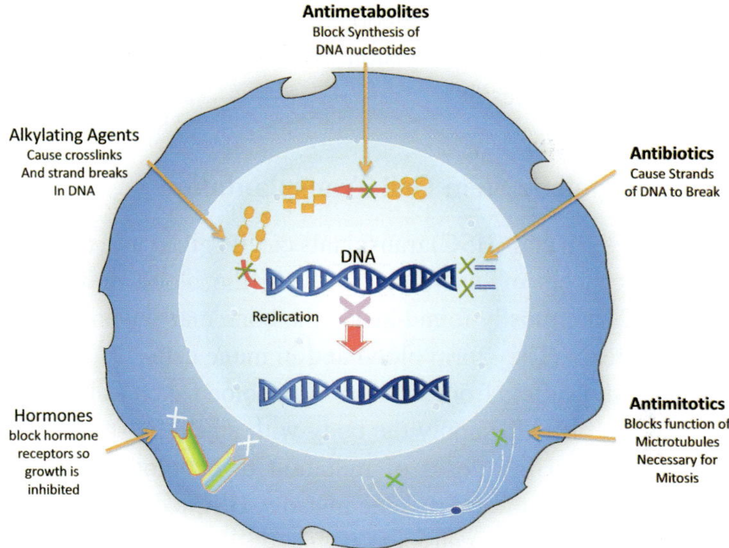

Figure 2.4   Common chemotherapy agents.

2.10.1.3. *Mitotic inhibitors*

These are drugs that prevent the cell from dividing by disrupting the microtubule from being formed, thus preventing the cell from being able to pull apart into two separate cells.[43] A potential side effect of mitotic inhibitors is that they can also prevent nerve fiber formation that can lead to peripheral nerve injury and numbness.[44]

2.10.1.4. *Topoisomerase inhibitors*

These are substances that block topoisomerases. A topoisomerase is an enzyme that breaks and reconnects strands of DNA.[45] If it is inhibited, cellular division can also be prevented.

2.10.1.5. *Anthracyclines*

Anthracyclines are a class of antitumor antibiotics that are derived from *Streptomyces*, a bacterium found in soil.[46] While the mechanisms of anthracyclines are not fully understood, it is known that DNA replication is interrupted, thus preventing division. However, a common side effect is that cardiotoxicity can result.

2.11. Hematopoietic Stem Cell Transplantation

Hematopoietic stem cell (HSC) transplants can be given as a treatment for patients with cancers of disorders involving the blood or the immune system.[47] HSCs are mostly found inside the bone marrow and sometimes in the blood stream. They form blood and immune cells.

A patient who suffers from a cancer of the blood such as leukemia can receive HSCs in order to supply the body with a healthy supply of blood and immune cells. High doses of radiation or chemotherapy can also cause detrimental side effects to bone marrow. HSC's can be transplanted as a treatment due to those causes as well.

The most common form of transplant comes from the patient themselves. This is referred to as an **autologous transplant**.[48] Other forms of donors are listed as follows:

- **Allogeneic transplant** — From a related or unrelated donor.
- **Syngeneic transplant** — From an identical twin.
- **Cord blood** — From the umbilical cord of a donor or self.

An autologous transplant involves four phases. First chemotherapy is administered to the donor in order to reduce disease. They are then given growth factors to encourage proliferation of HSCs. The blood is then harvested from the donor by aphaeresis. The recipient undergoes a procedure to "wipe out" the immune system and bone marrow in preparation for the donation. A common method for this is Total Body Irradiation (TBI). The final phase is the engraftment phase, in which the stem cells are given to the recipient.

Graft-versus-host disease (GVHD) is a complication that can follow a transplant. It is a result of the body mounting an immune defense against the new tissue.[49] The usual cause is that the body's T-cells identify the donation as foreign and act to destroy it. As a means of preventing this from occurring, it is common treatment to suppress the immune response. It can be suppressed with corticosteroids or antibodies to T-cells can be given. As a result of the suppression of the immune system, another negative aspect is that the recipient will be more prone to an infection, because the body would not be able to fight as many illnesses on its own.

An acute case of GVHD usually occurs within the first three months, whereas chronic GVHD would usually occur after three months and could potentially last for the rest of the recipient's life. Acute GVHD could be helpful in the aspect that transplanted immune cells could mount an attack against remaining cancer cells.

2.12. Hormonal Therapy

Hormone therapy works against hormone receptors for sex hormones. For instance, an antiestrogen can be administered for some forms of breast cancer.[50,51] An antiestrogen can either reduce the amount of estrogen produced or block its receptor site on the tumor. Other forms of hormone therapy exist, such as aromatase inhibitors. Aromatase inhibitors work by blocking the enzyme aromatase.[52] Aromatase helps convert androgens into estrogen. However, aromatase inhibitors are only effective when treating postmenopausal women because aromatase inhibitors cannot prevent the ovaries from producing estrogen.

Men with prostate cancer can benefit from hormone therapy. Male sex hormones can cause cancer growth in the prostate. An agonist of the luteinizing hormone (LH) or antiandrogen can be given to prevent testosterone from being produced by the testes, a drug could be administered to prevent the hormone from being able to enter the tumor or the testicles could be removed.

2.13. Biological Therapies

The body's own defenses can be used to fight cancer. Research has developed methods in which the body's immune system can be enhanced, mimicked, or manipulated in order to destroy cancer. Scientists have been able to develop drugs that manipulate molecular pathways and genes.[13]

2.13.1. *Types of biological therapies*

2.13.1.1. *Monoclonal antibodies*

Monoclonal antibodies (MAbs) are specific types of antibody that are produced in a laboratory setting.[53] Original research utilized mouse antibodies; however, advancements in understanding MAbs allow for them to be made with human proteins incorporated in with mouse proteins, or entirely made from human proteins. The addition of human proteins is more effective because the body does not mount an immune response as readily as it would with mouse proteins. Many forms of MAbs have been developed and approved as targeting agents for various forms of cancer.

Rituxan® is cancer drug that utilized MAbs and has been approved by the FDA to treat some forms of cancers such as non-Hodgkin lymphoma and some forms of leukemia.[54] Many other drugs have also used this technology to target other forms of cancer. Some even use radioactive antibody-based drugs to deliver radiation directly to the cancer cells.

2.13.1.2. *Angiogenesis inhibitors*

These are drugs that are found naturally or synthetically made that inhibit the ability of a tumor to produce more blood vessels.[55] Normally when a tumor is growing and dividing, it increases the number of blood vessels going through it, in order to deliver more nutrients to the growing tumor. Tumors cannot grow without the formation of new blood vessels; therefore, angiogenesis inhibitors can prevent or slow the growth of tumors.

The drug Avastin® is an angiogenesis inhibitor, which utilizes monoclonial antibodies and has been approved by the Food and Drug Administration (FDA).[56] It is often used with other drugs to treat some forms of cancers that have metastasized.

While angiogenesis inhibitors are not very effective by themselves, they are beneficial in preventing the growth of a tumor. An added benefit is that there is little toxicity to the body and tumors do not develop immunity to angiogenesis inhibitors given over a long period of time. However, if delivered over a longer period of time, problems with bleeding, poor wound healing, and blood clotting are possible side effects.

2.13.1.3. *Tyrosine kinase inhibitors*

Tyrosine kinase inhibitors (TKIs) are a class of drugs that block enzymes, called tyrosine kinases, which are responsible for cell growth and division.[57] A TKI drug called Imatinib (Gleevec) can treat people with chronic myeloid leukemia, a stomach cancer called GIST, dermatofibrosarcoma protuberans, and acute lymphoblastic leukemia. Other forms of leukemia, which Gleevec cannot treat, are often treatable with other TKIs. For instance, dasatinib (Sprycel) and nilotinib (Tasigna) can treat many leukemias that are resistant to treatment with Gleevec.

2.13.1.4. *Proteasome inhibitors*

These are drugs that prevent the proteins from being broken down inside of a cell.[57] This causes a buildup of proteins within the cell that can lead to cell death. A common form of proteasome inhibitor is the drug Botezomib (Velcade), which is used to treat myeloma.

2.13.1.5 *mTOR inhibitors*

These block mammalian targets of rapamycin (mTOR), a type of serine/threonine protein kinase.[58] This kinase regulates cellular growth, motility, and survival. If it is inhibited, translation of genes involved in the cell cycle, in particular genes involved in the development of new blood vessels, is blocked. Two FDA-approved mTOR inhibitors are Torisel®[59] and Afinitor®,[60] both of which act to treat advanced renal cell carcinomas.

2.13.1.6. *Immunotherapy*

It is a type of cancer therapy that uses the body's own immune system to target and fight cancer.[60] Immunotherapy can work by two main methods. It can either boost the patient's immune system (active immunotherapy) or an immune system component, such as antibodies (passive immunotherapy), made in a laboratory can be given. It can be used to target a specific type of cell (specific immunotherapy) or generally stimulate the entire immune system (nonspecific immunotherapy). Immunotherapy can be used independently or with other treatments. Initially research in immunotherapy, as a method for cancer treatment, originated due to a correlation between infections following surgery having a positive impact on cancer patients. Since the concept involves using a patient's own immune system, immunotherapeutic treatment is a less toxic form of treatment. By enhancing the immune system, the body's ability to recognize cancer cells as foreign is also enhanced.

Immunotherapy enhances the body's immune system by stimulating cytokines, which in turn can stimulate the production of antibodies and T-cells that can decrease the growth and spread of tumors.

Interferons are a type of cytokine that the body naturally produces and can be readily synthesized in the laboratory setting.[61] Interferon-alpha is a type of cytokine that has been approved by the FDA to slow the tumor growth for multiple myeloma, some forms of leukemia and melanoma. Interleukins are another type of cytokine that have also been used to treat kidney cancer and melanoma. Possible side effects of receiving cytokines to suppress tumor growth are general feeling of discomfort and flu-like symptoms.

The use of vaccines to prevent cancer is currently being investigated and undergoing clinical trials. By administering a vaccination as a prevention of cancer, the body is more able to recognize and attack cancer cell before a tumor can grow. A major benefit of vaccinations is that no normal tissue will be damaged, only the cancerous cells.

2.14. Clinical Trials

After extensive research has been performed, one of the final stages is the clinical trial.[62] Clinical trials can help improve treatment or detection methods and decrease the amount of harmful side effects. Undergoing a clinical trial helps scientists determine whether they are able to improve methods of treatment and detection. However, possible problems are not always known, so there is a level of risk involved when one agrees to participate. There are various types of clinical trials; they can focus on prevention, diagnosis, screening, or improving quality of life. Most clinical trials are classified into one of the three or four phases.

Phase I usually evaluates the method in which a drug should be delivered (mouth, injection into the bloodstream, injection into the muscle, etc.), how frequently the drug should be given and the dose. The side effects are also noted in this stage. Usually Phase I involves only a small group of patients, sometimes as few as a dozen.

Phase II continues to test the safety of the drug and analyze how well it works. Phase II also tends to focus on specific types of cancer.

Phase III compares the experimental drug or combinations of drugs to current standards. Patients are usually randomly assigned a treatment, in order to obtain more meaningful results. Phase III usually involves thousands of patients.

Phase IV is performed after treatment is approved and the drug is already on the market. It is usually used to examine side effects, risks, and benefits over a long period of time. This phase involves using thousands of patients (usually more than the number analyzed for Phase III).

Before patients enroll in a clinical trial, they are asked to learn about their disease and the purpose of the clinical trial, and speak with a professional (such as their physician). Once they have decided that they would like to participate in the trial, they are requested to sign an informed consent form. The consent form educates the participant about the key facts, purposes, possible risks and benefits, and tests that they might need to undergo. By signing the consent form, the patient agrees to undergo the trial, but does not waive their legal rights. Throughout the process, communication between the patient and the research team and health professionals is maintained, so that the patient remains informed about all of their actions.

2.15. Long-Term Survival

Many patients, when receiving news of cancer, are given information from their physician explaining their diagnosis, treatment options, and a survivorship care plan that includes information on long-term and late effects following treatment.[63] Once a patient completes their treatment regiment, they often experience side effects or new problems associated with their treatment. There is an increasing number of survivorship centers in the United States and around the world that are compiling information, in order to better help patients become informed on what to expect and how to care for themselves during and following treatment. Currently, long-term effects are poorly documented and understood, so these centers are working on compiling data, in order to improve long-term patient care.

Currently, a large body of information is available about the long-term effects of pediatric cancers due to the Childhood Cancer Survivor Study which was started in the year 1993.[64] This study has monitored children, who had cancer and had received treatment, over the majority of their lives. The study was able to determine that of all of the children involved in the study, two-thirds of the children had at least one form of a chronic health condition and one-third had more than one problem. This study focuses on childhood cancers, but gives much insight, which indicates that most cancer survivors struggle with complications throughout the rest of their lives. Since more than 60% of cancer survivors are 65 years or older, many more factors must be analyzed in order to determine whether a complication resulted due to treatment, aging, or both.

A long-term effect following cancer treatment usually starts during the treatment and could persist for months or years following the treatment. A late effect would happen well after the administration of therapy. Examples of late effects of cancer could be the development of secondary cancers such as blood cancers or solid tumors, possible due to radiation or certain chemotherapeutic agents causing DNA damage. Lymphedema is also sometimes a late effect. It is characterized by chronic swelling due to buildup of fluid. Lymphedema is usually found in one-fifth of patients who had lymph nodes removed or damaged during treatment. Currently there is no cure for lymphedema, but it can be managed and treated once identified.

2.16. Prevention

While many forms of treatment exist for treating cancer and research is continuously improving treatment and diagnostic methods, the best approach is to prevent the cancer from ever occurring in the first place.[65] There are simple choices and steps that can be made, on a regular basis, in order to decrease the risk of developing cancer.

Smoking is a leading cause of lung and oral cancers.[66] Smoking is also the leading cause of preventable premature deaths that occur in the United States. There are over 250 harmful chemicals contained within a cigarette and 50 of them have been confirmed to be carcinogenic. Not only does smoking cause lung and oral cancers, it has also been linked to bladder, cervical, esophageal, kidney, lip, pancreatic, voice box, and throat cancers. It is not only harmful to the person smoking but also to those around that person. Passive smoking also puts a person at risk for developing these cancers. Other forms of tobacco, such as chewing tobacco are also linked to esophageal, mouth, pancreatic, and throat cancers.

Diet plays a very large role in the health of a person. A diet high in saturated fat versus a diet that contains a mixture of fruits, vegetables, whole grains, and low fat show drastically different results. By eating a mixture of fruits, vegetables, and whole grains, the risks of developing colon, esophageal, lung, and stomach cancers drastically decrease. An added benefit of having a healthy diet is that a healthy diet usually correlates to a healthy body weight, which also decreases the risk of colon, breast, prostate, and uterine cancers.

Alcohol consumption should be in moderation. The regularity and the amount of alcohol consumed can influence the risk of developing cancers of the mouth, throat, esophagus, kidney, liver, and breast. If a man were to consume only two drinks a day (one drink a day for a woman), the risk of developing these forms of cancer is still elevated.

Maintaining a healthy weight through diet and regular physical activity will drastically decrease the occurrence of colon, breast, prostate, and uterine cancers.

Sun exposure continues to be the leading cause of skin cancers. It is important to protect one's skin by either avoiding the peak radiation hours (10 a.m.–4 p.m.) or use sunscreen (SPF 15–30) while outdoors. Tanning beds and sunlamps have an amount of risk determined by the extent of exposure associated with them.

Immunizations can be given to prevent viral infections that have been linked to cancer. The hepatitis B vaccination is now given to infants in order to help prevent a condition of liver inflammation that can lead to liver cancer. The Gardasil vaccine is also given to females up until the age of 26 years, in order to prevent a virus that can cause cervical cancer.[20]

Avoiding risky behavior such as unprotected sex with casual partners and making sure that one's hands and food are properly cleaned can drastically reduce the chance of contracting HPV, HIV, and hepatitis B and C.

Getting screened on a regular basis can help through an early detection of cancers. Having regular female examinations, prostate exams, breast exams, and colonoscopies can aid in early detection and prevention of several common cancers. If a cancer is caught at an early stage, it is more easily treated than after it has spread.

Chapter

3

Principles of Neutron
Capture Therapy

3.1. Boron Neutron Capture Therapy (BNCT)

Boron Neutron Capture Therapy (BNCT) is a binary cancer treatment in which compounds containing ^{10}B are selectively introduced into tumor cells and then irradiated with thermal neutrons. The ensuing sequence of events is given in Fig. 3.1(a) and summarized graphically in Fig. 3.1(b). The attractiveness of BNCT lies in the fact that the neutron-capture cross section of boron is more than three orders of magnitude higher than for other nuclei common to living tissue, the mechanism for BNCT tumor cells eradication is pictured in Fig. 3.1(b). There are other isotopes that have large neutron capture radii, such as ^{235}U ($\sigma_{th} = 681$ b) and ^{6}Li ($\sigma_{th} = 941$ b), but few greater than ^{10}B ($\sigma_{th} = 3838$ b). Boron also has the advantages that the element itself is not radioactive and it is fairly nontoxic. Boronic groups can be incorporated into compounds having hydrolytically stable linkages between boron and other elements such as C, O and N. The boron concentration needed to be effective in BNCT is generally estimated at 10^9 ^{10}B atoms (natural abundance 19.9%) per cell, which translates to approximately 35 μg ^{10}B/g of tissue.[11b] Under these conditions, it has been estimated that approximately 85% of the radiation damage arises from the neutron capture reaction. To prevent damage to healthy tissue in the path of the neutron beam, the surrounding tissue

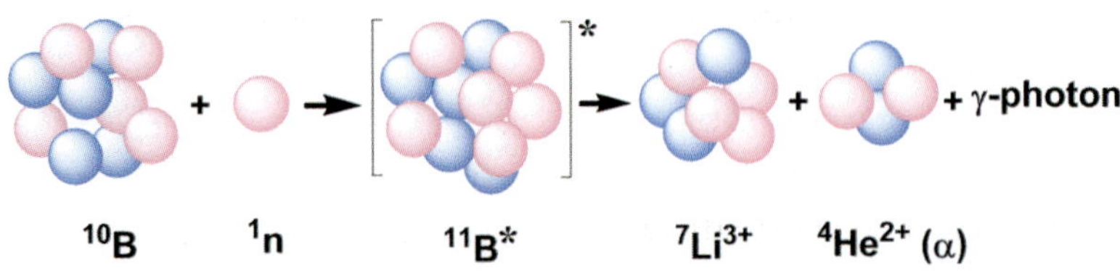

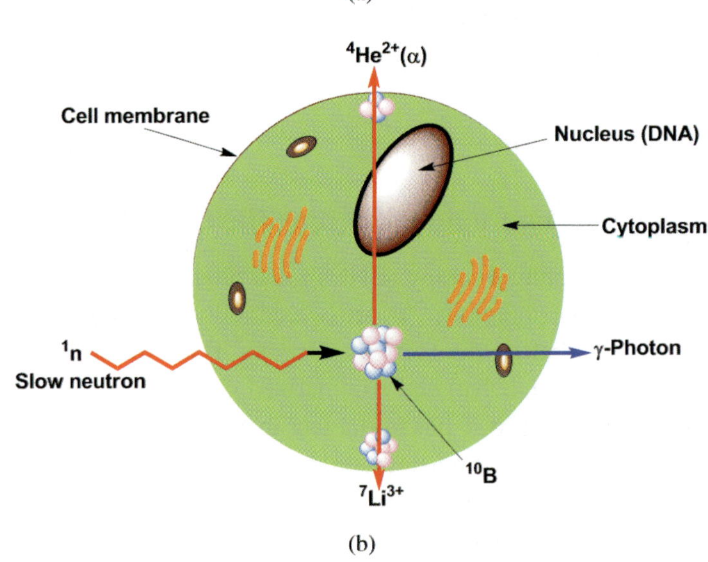

Figure 3.1 BNCT killing mechanism.

should contain no more than 5 μg of ^{10}B/g of tissue. This rather large amount of ^{10}B needed in a tumor cell puts an increased burden on boron delivery agents that is not found in other forms of chemotherapeutic compounds.

A primary reason for the development of BNCT is in the treatment of malignant brain tumors, such as glioblastoma multiforme (GBM). This is one of the most malignant forms of cancer that infiltrates the brain so aggressively that surgery is rarely able to remove all the cancerous tissue; it is virtually untreatable and is inevitably lethal.[67,68] On the other hand, since these types of tumors have little history of metastasizing to other organ sites in the body, if tumor eradication could be achieved in the brain, a significant increase in life expectancy could be obtained. There is an additional complication in the case of brain tumors in that the therapeutic compounds must be able to traverse the blood brain barrier (BBB) in order to reach infiltrating cancer cells. Although the tumor itself may have an immature, leaky, vascular system that would allow the entry of BNCT agents, it is the eradication of the cancer cells that might have infiltrated healthy brain tissue, and are still protected by an intact BBB, that ultimately determines the long-term survival of patients.

3.2. Gadolinium Neutron Capture Therapy (GdNCT)

Figure 3.2 schematically shows the ^{157}Gd(n, γ)^{158}Gd reaction. The ^{157}Gd atom, which occurs in 15.6% natural abundance, releases Auger electrons, internal conversion (IC) electrons, γ-rays, and X-rays after thermal neutron capture across its wide capture cross section of 255,000 b, which is 65 times greater than the thermal-neutron-capture cross section of ^{10}B. Although Auger electrons travel miniscule distances of only several 100 Ångstrom, they are able to kill tumor cells if the ^{157}Gd is located in close proximity to DNA.[11a] Concomitant X-rays are negligibly lethal owing to their relatively low energy. Thus, γ rays, IC electrons, and Auger electrons are mainly responsible for the tumoricidal effect of Gadolinium Neutron Capture Therapy GdNCT. For comparison, the total kinetic energy of the various particles is 7.9 MeV, almost double the energy released by ^{10}B(n, α)^{7}Li.

To employ this modality of NCT, it is essential to have a source of well-thermalized neutrons. Suitable thermal neutron energy spectra can be obtained at the NCT facilities of the Kyoto University Research Reactor Institute (KUR) and the Japan Atomic Energy Research Institute (JRR4). A Maxwellian distribution, with good elimination of concomitant γ-rays from the irradiation field, can be achieved by scattering with Bi atoms (bismuth scatter). In these facilites, the Cd ratio, that is, the proportional fluence of thermal neutrons (<10 eV) and nonthermal neutrons (>10 eV), is very large (150), while the γ-ray contamination in the irradiation field is

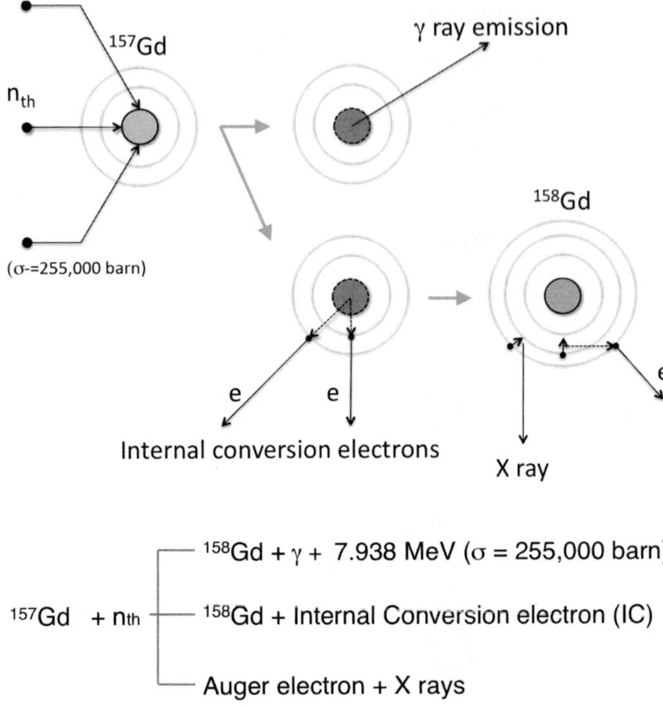

$$^{158}\text{Gd} + \gamma + 7.938 \text{ MeV } (\sigma = 255,000 \text{ barn})$$

$$^{157}\text{Gd} + n_{th} \quad ^{158}\text{Gd} + \text{Internal Conversion electron (IC)}$$

$$\text{Auger electron} + \text{X rays}$$

Figure 3.2 ^{157}Gd(n, g)^{158}Gd neutron capture reaction.

comparatively small (5×10^{-3} Sv/n). These facilities can also yield epithermal neutron fields with a Cd ratio adjustable from 150 to 9.4, depending on how the D_2O filter is set, for example, filtering with a boral[11b] plate-containing cadmium. Generally, thermal neutron beams are applied to superficial tumors, such as cutaneous melanoma, while epithermal beams are more effective for brain tumors.

Why gadolinium? As Table 3.1 shows the most promising nuclides for NCT are ^{10}B and ^{157}Gd atoms, which have stable isotopes and large thermal-neutron-capture cross sections. However, free Gd^{3+} is toxic and so must be chemically stabilized by chelation.

The ^{157}Gd atom captures neutrons across a very large cross section of 255,000 b. One can compare the ^{157}Gd atom to the Thousand-Armed Goddess of Mercy (see Fig. 3.3), or Kannon, a manifestation of the Buddha who accepts prayers and is believed to have the power to grant

Table 3.1 Thermal-neutron-capture cross section (σ) of nuclide candidates for NCT.

Nuclide	Cross Section (σ)	Nuclide	Cross Section (σ)
^{135}Xe*	2,720,000	^{3}He	5500
^{157}Gd	240,000	^{10}B	3837
^{158}Gd	58,000	^{199}Hg	2000
^{149}Sm	41,500	^{241}Pu*	1375
^{113}Cd	20,000	^{6}Li	953
^{242}Am*	8000	^{235}U	678
^{151}Eu	5900	^{174}Hf	400

Figure 3.3 The Thousand-Armed Goddess of Mercy (Kannon) at Kyoto Sanjusangendo.

anyone the blessing of complete relief. If we scale the diameter of the ^{157}Gd nucleus to 1 cm, the nominal 2-km width of the outstretched hands would comfortably capture any passing thermal neutrons.

High-LET Auger electrons cause lethal damage to cells by breaking double-stranded DNA (see Fig. 3.4). However, because of the limited trajectory of the Auger electrons, the ^{157}Gd atoms must be in close proximity to the DNA to be effective. While rupture of both strands is lethal to tumor cells, single strand breakage is reparable via DNA polymerase and usually sublethal. Thus, the more Gd atoms that can be positioned close to the tumor DNA by ligand binding, the more potent

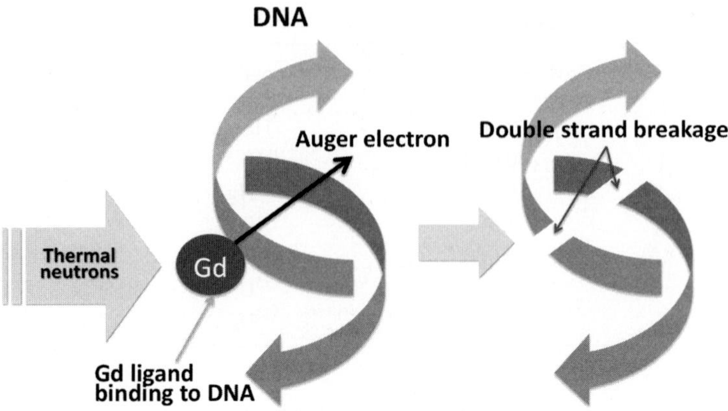

Figure 3.4   Lethal damage to double-strand DNA.

will be the killing effect from double-strand DNA breakage.[11a] High lethality cannot be expected when Gd atoms are mainly located in interstitial spaces or cytoplasm or both.

Chapter 4

Major Neutron Capture Therapy (NCT) Drug Prototypes

4.1. BNCT Agents: Background and General Requirements

One of the first experiments on the use of BNCT in cancer therapy was initiated in 1938 using the cyclotron at the University of Illinois to generate the neutrons. It was later moved to the University of California at Berkeley to take advantage of a higher energy neutron beam.[69] Samples of a mammary carcinoma, a lymphoma and an undifferentiated sarcoma were grown in mice and then removed after about 10 days growth, infused with boronic acid (H_3BO_3), irradiated with neutrons and then reimplanted into the test animals. The results showed that tissues treated with boron and irradiated with slow neutrons tended not to regenerate. This was one of the first demonstrations that neoplastic cells could be destroyed *in vivo* if sufficient boron was delivered to the cells and then irradiated with thermal neutrons.[69] It was not until the completion of the Brookhaven Graphite Research Reactor (BGRR) in 1950 that research in BNCT was undertaken again.

In 1951, a nuclear reactor was established for medical purposes at Brookhaven National Laboratory in Long Island (see Fig. 4.1); this was the first one designed for BNCT. Nineteen malignant brain tumors were treated there from 1953 to 1959. Dr William H. Sweet (1911–2001), a neurosurgeon of Massachusetts General Hospital promoted this project. He studied physics and medicine in Harvard University. He became

41

Figure 4.1 Nuclear reactor for medical purpose, Brookhaven National Laboratory. Center cube: reactor: *left*: hot laboratory; *right circles*: wards.

Source: Hatanaka, H., Slow neutron therapy. *Radiol. Med.* 4, **1984**, 83–106.

Professor of Surgery at Harvard Medical School and a chief of the Neurological Service at Massachusetts General Hospital. He also treated 17 cases at the research reactor of Massachusetts Institute of Technology (MIT) in Boston during 1960–1961. In MIT, he had an operating theater in a basement of the reactor. He died of pneumonia on January 22, 2001, at his home in Brooklyn.

From 1951–1961, early clinical trials of BNCT were conducted at the Brookhaven National Laboratory and at the Massachusetts General Hospital using thermal neutrons and sodium tetraborate (borax) as the ^{10}B vehicle. The first patient of a 10-patient trail was irradiated in 1951, shortly after the reactor was commissioned. The patients had undergone neurosurgical removal of malignant cerebral glioma but all showed signs of recurrence. The irradiation time was from 17–40 min. There were no serious side effects from BNCT and the median survival time (MST) was 97 days. A second group of nine malignant glioma patients were irradiated with a higher neutron flux and a higher ^{10}B concentration. Although the MST was 147 days, many patients developed severe, intractable radiation damage of the scalp.[70] However, because of insufficient neutron beam penetration, these BNCT trials demonstrated neither significant prolongation of survival nor evidence of therapeutic efficacy.[70] Because of the disappointing results, clinical trails in the United States were suspended. BNCT research was transferred to Japan. Dr Hiroshi Hatanaka (1932–1994), Neurosurgeon of Tokyo University, had studied BNCT with Professor Sweet. He was a devout Protestant and had a thorough knowledge of Confucianism. He continued BNCT research with Professor Keiji Sano, Professor of Neurosurgery of Tokyo University, in Japan and they treated the first patient with thermalized neutrons from the Hitachi Nuclear Reactor using BSH (sodium salt of mercaptoundecahydrododecaborate, $Na_2B_{12}H_{11}SH$) modified by Shionogi Pharmaceutical Industry in 1968. Unfortunately the patient died 4 months after BNCT, but their first trial was an event of great historical significance. After 20 years experiences, in 1998, Professor Hatanaka, Teikyo University, reported a good outcome on 101 cases using various nuclear reactors of the Musashi Institute of Technology, the Japan Atomic Research Agency, and the Research Reactor Institute of Kyoto University.

In the 1980s, a second clinical BNCT trial was done on malignant melanoma using BPA (*L*-4-dihydroxyborylphenylalanine) that had been investigated as menanogenesis precursors by Yutaka Mishima, Professor of Dermatology at Kobe University.[73] Those clinical studies were sufficiency successful that BNCT studies were again undertaken worldwide.

In later tests, more penetrating epithermal neutron beams were used to reduce scalp damage without the complications of craniotomy. At the Massachusetts Institute of Technology Research Reactor II, epithermal neutron beam irradiation was first used in the BNCT treatment of human peripheral melanomas in 1994.[71] Meanwhile, at the Brookhaven National Laboratory, an epithermal neutron beam was used for the first time in a BNCT trial for the treatment of central nervous system (CNS) tumors.

Epithermal neutrons beams are now standard in all clinical trails of brain tumors. They are thermalized in the scalp or at the brain surface, the thermal neutron dose reaches its maximum intensity at a depth of ~2 cm, almost equal to the total thickness of scalp and skull (Fig. 4.2). Thus, thermal neutron doses reach their highest concentration at the brain surface when the epithermal neutrons are irradiated onto the scalp surface. This modality

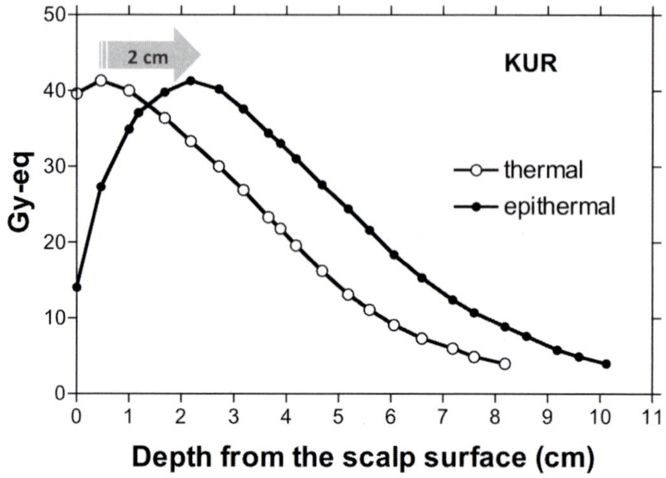

Figure 4.2   Depth distribution of neutrons in BNCT.

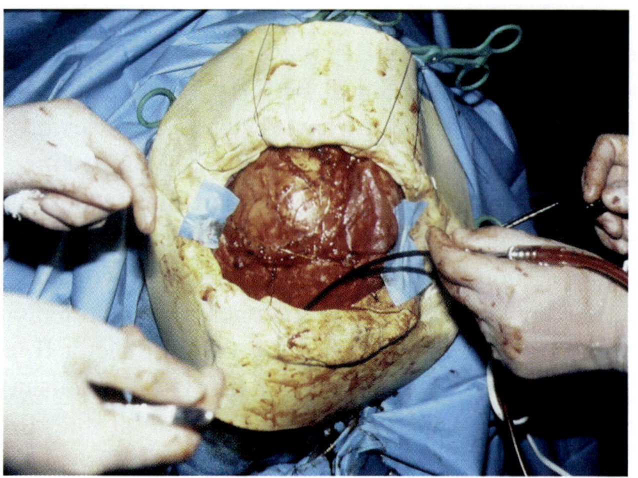

Figure 4.3   Early BNCT of brain tumors.

allows BNCT without craniotomy. In the early BNCT protocols, the scalp and the skull had to be opened in order to make thermal neutron penetrate deeply (see Fig 4.3).

The head was protected from the thermal neutron bombardment via flexible lithium sheet (yellow in figure). In the current BNCT protocol,

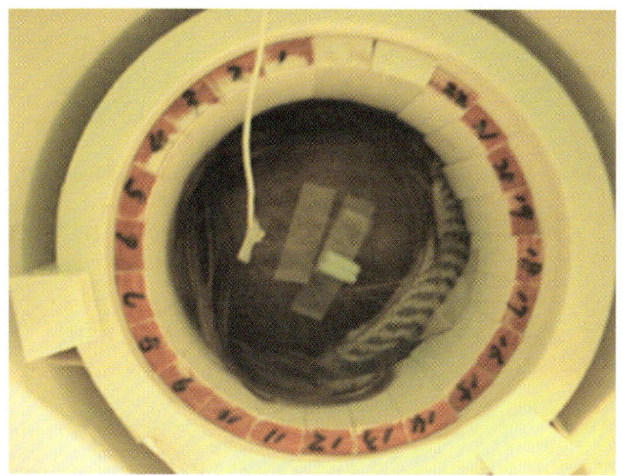

Figure 4.4 Current BNCT of brain tumors.

epithermal neutrons are irradiated onto the scalp without any surgical maneuvers (Fig. 4.4).

In addition to treating difficult brain tumors as discussed above, BNCT has been successfully employed in other cancer therapies. For example, a Finnish research group recently reported that a patient with recurrent head and neck tumors showed an excellent response to BNCT treatments.[72] In Japan, BNCT has been employed to treat cutaneous malignant melanoma. Clinical tumor regressions were demonstrated in response to BNCT.[73] In the period 1997–2000, BNCT clinical studies were conducted in Argentina, Czech Republic, Finland, Japan, Netherlands, and Sweden using epithermal neutrons.[74] Currently, BNCT is still being optimized and evaluated for safety and efficacy in the USA, Japan, and Europe. At present, BNCT should be considered as an investigational treatment. The patients should be limited to conducting BNCT Phase I or Phase II clinical trials. BNCT is a complex binary cancer treatment, in which interdisciplinary interaction amongst scientists in many fields is indispensable. However, the successful delivery of ^{10}B containing compounds in cancer cells, as opposed to normal tissue is a critical difficulty facing BNCT.

Enhancing selectivity has been an ongoing challenge to chemists developing newer boron delivery agents. Emphasis has been on improving the

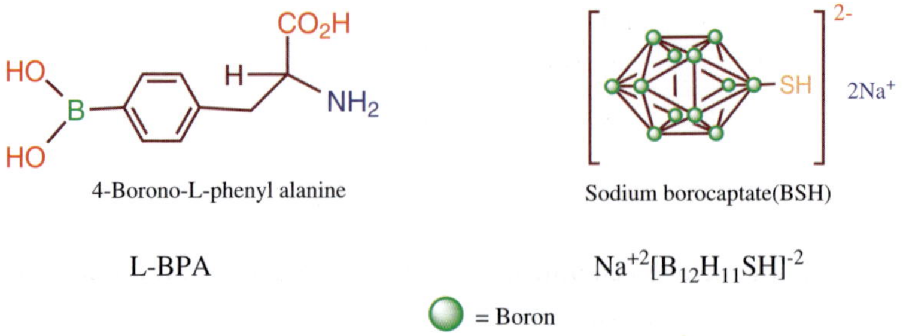

4-Borono-L-phenyl alanine

L-BPA

Sodium borocaptate(BSH)

$Na^{+2}[B_{12}H_{11}SH]^{-2}$

○ = Boron

Figure 4.5   Molecular structures of BPA and BSH.

concentration ratios of tumor-to-blood and tumor-to-normal tissue. In the last decades, a variety of carrier molecules have been investigated as possible boron delivery agents. These include carbohydrates[75] polyamines,[76] nucleosides,[77] antibodies,[78] porphyrins,[79] liposomes[80] and amino acids.[81] Many of these will be discussed in more detail throughout this book. Despite these efforts, only two low molecular weight compounds, BPA and BSH (Fig. 4.5) are currently used in clinical trials. The results on these two boron compounds are not universally promising, owing to their low tumor-to-blood and tumor-to-brain tissue ^{10}B ratios.[82–88] Therefore, it is important to develop additional potential precursor molecules and more sophisticated delivery systems in BNCT. A promising trend has been in the development of nanomaterials, such as liposomes, dendrimers, nanotubes, and magnetic nanoparticles, as both boron host molecules and delivery agents.[89] These developments will be detailed in following chapters.

Presently, epithermal neutron beams, which may penetrate a distance of around 10 cm, are commonly used in BNCT clinical treatments. As these neutrons penetrate through tissue, they must slow down to the thermal range when they reach the tumor. Therefore, the beam strength must be calibrated in terms of tumor depth and the nature of the tissue that must be traversed. Neutron beams currently used are composed of multiple beams such as epithermal, thermal, fast neutrons, and γ-rays. The latter two contribute to background radiation and should be filtered out of the beam. Even with a properly filtered beam, nuclear capture reactions will occur within tissue cells which cannot be filtered out and contribute to the background radiation. The most important are nitrogen capture reactions (Eq. (4.1)) and proton capture reactions (Eq. (4.2)).[91]

$$^{14}N + n_{th} \rightarrow [^{15}N]^* \rightarrow {}^{14}C + {}^1H \qquad (4.1)$$

$$^1H + n_{th} \rightarrow {}^2D + \gamma \qquad (4.2)$$

Even though the neutron capture cross section of ^{10}B is several orders of magnitude higher than for most elements, the high tissue concentrations of some elements, such as hydrogen and nitrogen, in normal tissue (see Table 1.1 in Chap. 1) are such that their neutron capture contributes significantly to the background radiation dose. To avoid the deleterious side effects and reduce the background dose, for a successful BNCT clinical treatment, both the boron dose and the neutron dose must be carefully analyzed. This constitutes an important challenge in both boron drug delivery and neutron beam design.

In the coming years, high quality neutron beams will become more available, with the anticipation of the development of nuclear accelerators that can be housed in hospitals. However, the application of this development will depend on the prior availability of suitable boron delivery agents. Compared with other conventional chemotherapy, BNCT treatments would have the following advantages: (1) since BNCT is a binary method, based on nuclear fission of ^{10}B atom, the delivery agents themselves could be nontoxic, in the absence of neutron irradiation and of high ^{10}B content; (2) because the released energetic particles from the ^{10}B capture reaction are confined to the tumor cells, the damage to outside tissues should be very limited. As of this time, there is no one boron delivery agent that meets the criteria of high localization in tumor cells coupled with an otherwise low biological distribution. Since BNCT treatments were originally designed for brain tumors, such as GBM, that have the ability to infiltrate healthy brain tissue that is protected by the blood brain barrier (BBB), such agents must be able to traverse the BBB and seek out malignant neoplastic cells in the presence of healthy tissue.

4.1.1. *Blood brain barrier (BBB)*

The BBB is a monolayer of tightly packed endothelial cells that line the capillaries, venules, and arterioles of the brain that regulate the flow of molecules between the blood and the central nervous system (CNS) (see Fig 4.6). The

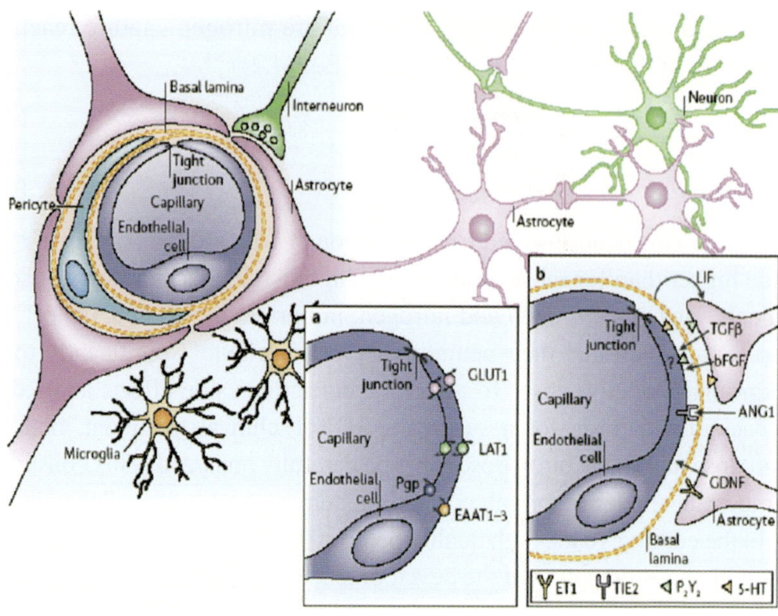

Figure 4.6   Diagram showing the cellular constituents of the (BBB).

Source: Reproduced with permission from Ref. 90c.

endothelial cells are surrounded by a basal lamina and astrocytic perivascular endfeet. The astrocytes provide a cellular link to the neutrons.

The problems in drug delivery to the CNS has been discussed in a recent review by Partridge,[90a] and much of the following discussion is taken from that review. It has been estimated that the total length of blood delivering capillaries in the human brain is ~600 to 700 km and the surface area of the capillary endothelium is ~20 m^2. Tight junctions (see Fig. 4.8) exist between these endothelial cells that prevent the passage of even small ions, so that the transendothelial electrical resistance is >1000 Ω cm^2 instead of the usual 2–20 Ω cm^2 found in peripheral capillaries.

Movement of molecules from blood to the brain must take place through the endothelium; there are no *para*-cellular pathways for solute molecules to access the brain from blood. The effectiveness of this barrier can be seen in Fig. 4.7, which shows a whole body autoradiogram of a mouse 30 min after an intravenous injection of radiolabeled histamine. As

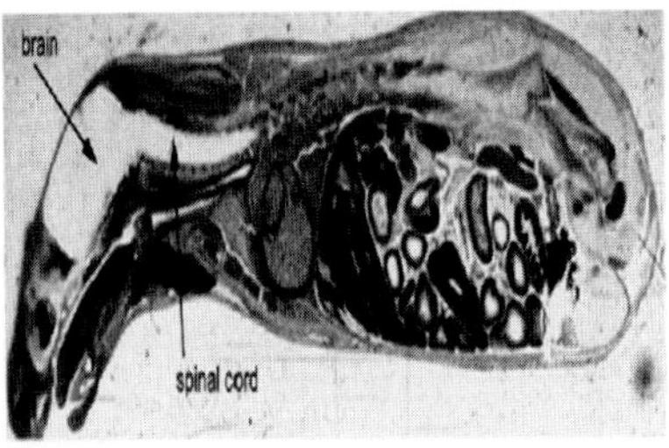

Figure 4.7 Whole body autoradiogram of an adult mouse sacrificed 30 min after injection with radiolabeled Histamine.[90a]

can be seen, the histamine traverses the more porous capillaries in the peripheral tissues, but is excluded from the CNS.[36]

The BBB is much more than a passive barrier since it regulates the passage of nutrients, metabolic fuels, peptides, and regulatory proteins. There are also active influx and efflux mechanisms (carrier-mediated transport (CMT) systems, active efflux transport (AET), and receptor-mediated transport (RMT) systems) by which the BBB regulates the passage of materials between the blood and the brain, and *vice versa*. Figure 4.8 summarized the basic modes of transport through the BBB.

In general, the BBB prevents the passage of drugs with molecular weights greater than 180 Da. In addition to being small, molecules that can traverse BBB cells are lipid-soluble, and have a low polar surface area (pathway b, Fig. 4.8). They also tend to have few potential hydrogen-bond-forming sites, including few, if any, carboxylic acid groups. Even molecules that have these characteristics are not assured of crossing the BBB by passive diffusion. For example, histamine, the compound involved in Fig. 4.8, has a mass of only 111 Da. It has been estimated that over 98% of small drug molecules cannot pass the BBB. It is also difficult to predict the effects of small changes in a molecule. Heroin penetrates the BBB about 100 times better than morphine; the only difference between the molecules is that morphine has two hydroxyl groups where in heroin they are converted to acetyl groups. This acetylation makes heroin much more lipid-soluble, allowing it to readily diffuse across the BBB.

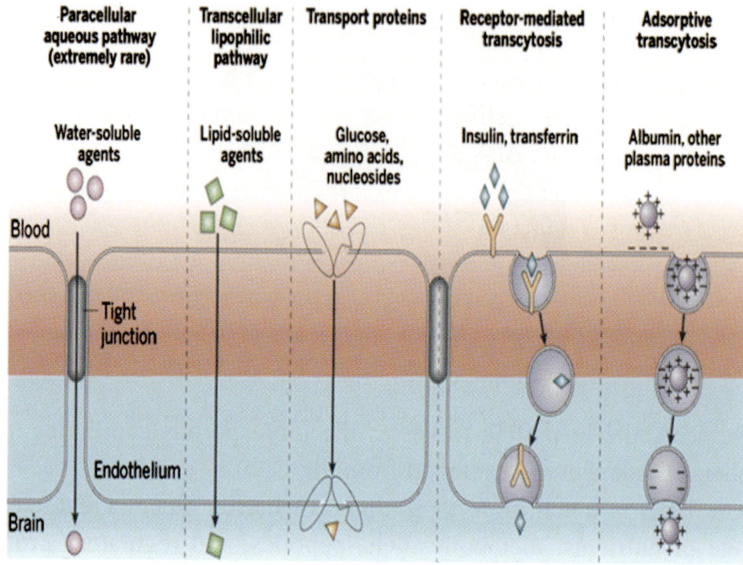

Figure 4.8   Pathways across the BBB.[90c]

Many of the new BNCT drugs are composites in which boron-containing moieties are attached to molecules needed for brain cells to function, these substances are passed through the BBB by CMT systems (pathways c–e, Fig. 4.8) There are some 298 transporter genes organized into 43 families[90b] that are used to transport building block and nutrient molecules across the BBB. For example, BPA traverses the BBB through such an active transport system. The development of both high- and low-molecular weight boron complexes with high preferences for cancer cells must be developed. It has been shown that the coadministration of BPA and BSH in C6-glioma bearing rats led to a tumor boron concentration which was much greater than that obtained from either compound alone and that normal brain boron concentration was only about one-sixth of that found in the tumor.[91] However, this does not automatically translate into more effective BNCT treatment. In a study of F98 glioma bearing mice, it was found that a combination of BPA and BSH, while leading to increased tumor boron concentrations, did not produce a MST any greater than that of BPA alone (33.5 days versus 33.4 days).[95]

4.1.2. *Drug delivery methods*

In addition to the development of tumor-specific boron carrier molecules, different modes of drug delivery have also been studied. Normally boron delivery drugs, such as BPA and BSH, are delivered intravenously (IV), while this is convenient, it might not be the best method for delivery of chemotherapeutic agents to brain tumors because of the BBB. Studies have shown that in patients with recurrent head and neck tumors, intra-arterial (IA) administration of BPA showed distinct advantages over intravenous administration.[92] However, Hatanaka and coworkers[93] studied the pharmacokinetics and boron uptake of BSH in 123 patients treated by BNCT from 1968 to 1994 for intracranial tumors with IA and IV infusion and found that, while IA administrated BSH can move more readily from blood to peripheral organs, there were no statistically significant differences in the tumor boron concentration and the tumor/blood ratio of boron concentration between the two techniques. More invasive approaches to drug delivery include disruption of the BBB,[94] the implantation of sustained release polymers,[95] direct intratumoral injection,[96] and convection-enhanced delivery (CED).[97] It was found that hyperosmotic mannitol-induced BBB disruption (BBB-D) and intracarotid infusion of BPA followed by BNCT in a rat brain model lead to a 246% increase in life span (ILS) (MST = 83 ± 41 days) compared to untreated controls (MST = 24 ± 3 days) and a 124% ILS increase over the IV injected animals (MST = 37 ± 3 days).[94] Similar results were obtained on the pharmacologic BBB disruption.[95] The development of enhance drug delivery systems that penetrate the BBB is critical to the long-term survival to patients with brain tumors.[96,97] However, an increased toxicity associated with BBB-D has also been found.[98] Physiologic techniques include the use of pseudo-nutrients such as cationic antibodies and chimeric peptides (those that are derived from genetically different zygotes).[99] In addition to disrupting the BBB, other methods such as intratumoral injection and CED have been investigated.[100] However, slow diffusion of a drug from any spot of entrance in the brain is a limitation. In CED, liquid is forced through the brain; however, results of CED on human patients with glioblastoma multiforme showed that the administered drug was restricted to a small area around the infusion sites.[100b]

Apparently, design and synthesis of new boron agents will be an important integral part of the preclinical and clinical evaluations of new capture agents for BNCT in the future. This chapter describes the potentially promising BNCT delivery agents including boron compounds and boron containing nanomaterials.

4.2. Boron Compounds-Based BNCT Agents

4.2.1. *Clinically used boron compounds*

BNCT delivery agents are required to meet the criteria of (1) high tumor uptake with high tumor/brain and high tumor/blood concentration ratios, (2) low systemic toxicity, and (3) rapid clearance from blood and normal tissues and persistence in tumor during BNCT. However, so far there is no single boron-delivery agent that fulfills all of these criteria. In clinical history, the first boron agents used in BNCT were sodium borates such as borax ($Na_2B_4O_7 \cdot 10H_2O$) and sodium pentaborate ($NaB_5O_8 \cdot 4H_2O$), as well as boric acid [$B(OH)_3$], which were chosen because of their low toxicity. These led to disappointing results because of the poor neutron beam source and only transiently high tumor/brain boron concentration ratios; the agents did not show any preference for neoplastic cells.[101–104] The development of synthetic routes to polyhedral borane anions (Fig. 4.9), such as $Na_2B_{10}H_{10}$ and $Na_2B_{12}H_{12}$ (see Refs. 105 and 106) were developed. These cage structures showed remarkable chemical, hydrolytic stabilities and low toxicity.[107] Thus, [10]B-enriched $Na_2B_{10}H_{10}$ was prepared and used in a clinical trial. However, the results of this clinical study were a total failure due to the high blood concentrations of the agents, which caused the vascular endothelium to be compromised during the radiation procedure.[108] There followed an intense search for compounds that preferentially localized in tumors. This resulted in the development of two boronated agents, L-4-dihydroxyborylphenylanaline (BPA, Scheme 4.1) and sodium mercapto-undecahydrododecaborate ($Na_2B_{12}H_{11}SH$, BSH; Fig. 4.9).

BPA was one of a group of arylboronic acids that showed low toxicity and favorable tumor/brain ratios. It can penetrate the BBB and has been used in the BNCT treatment of melanoma[109] and brain tumors.[110]

The drug is usually delivered by the IV injection of its fructose complex to improve water solubility. On the other hand, BSH was found to

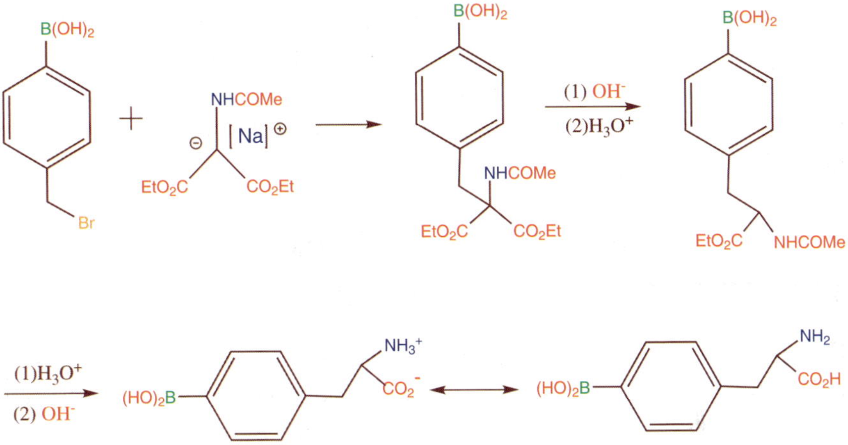

Scheme 4.1 Synthesis of BPA.

freely enter tumor cells but not cross the BBB. BSH was developed in response to the disappointing results from the clinical studies of the polyhedral boranes, such as $Na_2B_{12}H_{12}$, which were found to localize preferably in blood.[108] Comparative studies of $B_{12}H_{12}^{2-}$ and $B_{12}H_{11}SH^{2-}$ showed that, while BSH was more toxic than $B_{12}H_{12}^{2-}$ ($LD_{50} = 73 \pm 4$ mg B/Kg versus 1025 ± 15 mg B/KG in mice), it gave much more favorable tumor/blood ratios (1.4–20.0, average 8.0).[111] Beginning in 1968, clinical trails were conducted in Japan using BSH.[112] There is some question as to the fate of BSH under physiological conditions, the SH group can be easily oxidized to give the BSSB $(B_{24}H_{22}S_2)^{4-}$ dimer. Slow infusion of the monomer and dimer in melanoma-bearing mice and glioma-bearing rats showed that the dimer gave twice as much boron in the tumor, but also higher plasma and liver concentrations.[113] Analysis showed that the monomer was excreted in the urine of dimmer-infused rats. There is also evidence that the monomer, BSH, is altered under physiological conditions.[114]

While both BSH and BPA have limitations, a combination of the two drugs has been found to be more effective than any one of them alone.[115] A study of the microdistribution of ^{10}B after the infusion BPA and BSH found that BPA did not accumulate in quiescent (Q) cells, but only in

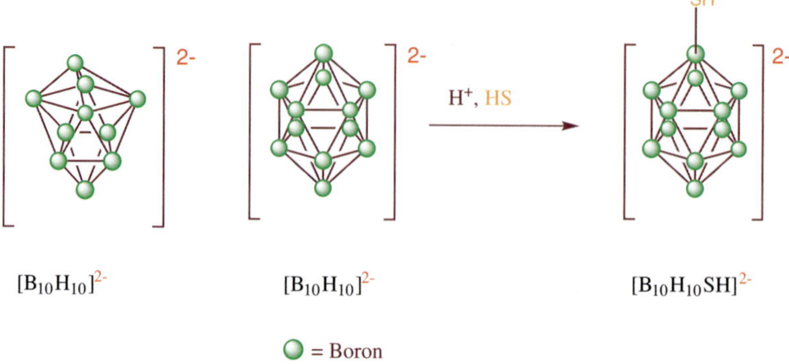

$[B_{10}H_{10}]^{2-}$ $[B_{10}H_{10}]^{2-}$ $[B_{10}H_{10}SH]^{2-}$

Figure 4.9   Selected polyhedral borane anions: decahydrodecaborate ($B_{10}H_{10}^{2-}$), dodecahydrododecaborate ($B_{12}H_{12}^{2-}$), mercaptoundecahydrododecaborate ($B_{12}H_{11}SH$)$^{2-}$.

actively dividing ones, whereas BSH entered into the Q cells much more efficiently. It has been suggested that combinations of drugs with complimentary properties may ultimately be the most therapeutically effective treatment modality.

Barth and coworkers[178] listed the requirements for a successful BNCT drug: (1) it must be nontoxic, with a rapid clearance from blood and normal tissues; (2) it must have a high tumor and low normal tissue uptake; (3) have a high tumor/blood and high tumor/normal tissue ratio; and (4) it must persist in the tumor long enough for radiation. Neither BPA nor BSH meet all of these criteria; thus, there is a continuing quest for better boron delivery agents. Some of these will be discussed in the following sections.

4.2.2. *Boronated polyamines as BNCT agents*

Amine boranes have been synthesized and their pharmacological behavior studied.[116] Part of the interest in these compounds is that the tetrahedral boron is isosteric to the carbon atom, but is more lipophilic. Amine boranes have been found to have, among other things, antineoplastic, and antiviral properties. One of the simplest neutral amine boranes, Me_3NBH_2COOH, was found to be a nontoxic, cytotoxic compound via multiple mechanisms. The drug achieved intracellular concentrations sufficient for BNCT, but not a tumor/blood ratio desirable for clinical applications.[116a] Although

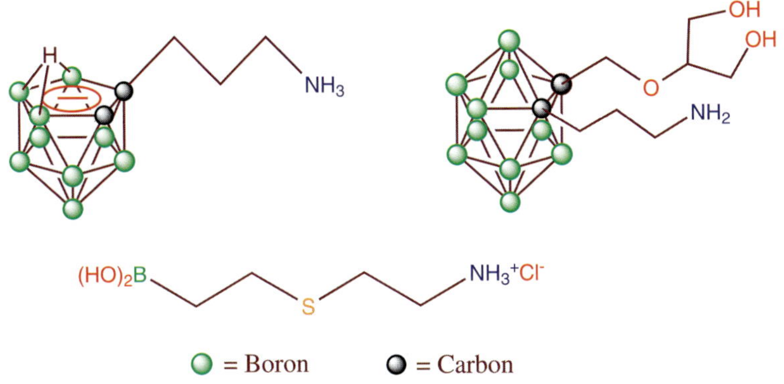

Figure 4.10   Boron-containing amines.

these compounds are not suitable as boron delivery agents for BNCT, they are important intermediates in the synthesis of new BNCT agents, some of which are shown in Fig. 4.10. The phenylalanine dipeptide of [Me_3NBH_2COOH] showed a tumor/blood ratio of 8:1; however, its effectiveness as a BNCT agent has not been explored.

A number of azanonaboranes-containing imidazole groups were synthesized (see Scheme 4.2) and their cytotoxicity to B16 melanoma cells were studied *in vitro*.[116b] Refluxing a 1:2 molar ratio mixture of histamine and [$(i\text{-}PrH_2N)B_8H_{11}NH(i\text{-}Pr)$] in benzene for 2 hr produced a mixture of the three products, 4, 5, 6 in Scheme 4.2, in yields of 30%, 42%, and 19%, respectively. The level of toxicity was a function of the amine that is attached to the azanonaborane. Compound 5, in which attachment was through the alkane -NH_2, was quite toxic even at the lowest concentrations used (13% cell survival at 50 μg B/mL), while compounds 4 and 6 were much less toxic (LC_{50} = 280 μg B/mL and 160 μg B/mL, respectively).[116b] The *in vitro* toxicities of the latter two compounds make them viable BNCT agents. Low toxicity was also found for the related compounds, [$H_2N(CH_2)_nH_2NB_8H_{11}NH(CH_2)_mNH_2$] ($n$ = 4, m = 4; n = 3, m = 4) in *in vitro* studies of V79 cells (Chinese hamster fibroblasts).[116c]

Polyamines, such as N-(3-aminopropyl)butane-1,4-diamine (spermidine, SPD) and N,N′-*bis*(3aminopropyl)butane-1,4-diamine (spermine SPM), are crucial feedstock to growing and proliferating cells, and their

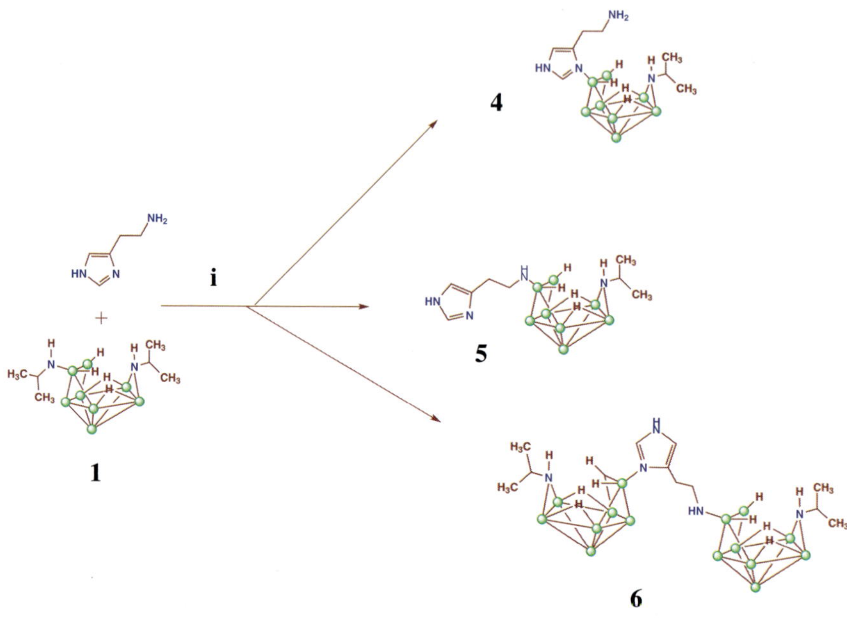

Scheme 4.2   Synthesis of imidazole/azanonaboranes.

absence has growth inhibitory effects on tumors.[117] Polyamines can accumulate inside cells in a high concentration of around 6.0 mM within a few hours, and they are important DNA active constituents; they bind to DNA through nonspecific electrostatic interactions. If BNCT agents can be delivered close to the DNA of tumor cells, the killing effectiveness should be dramatically improved. Assuming that the nucleus constitutes roughly 20% of the total cellular mass, DNA-localized BNCT agents would only have to deliver about 6 μg ^{10}B/g tissue. A number of carborane-attached analogues of SPD and SPM have been synthesized and tested *in vivo*, with disappointing results; they were either too toxic or did not accumulate in cancer cells.[118] Therefore, discovery of new DNA-binding agents remains a challenging target in BNCT research.[118,119]

Mono-substituted aminoalkyl and bis(long-chain-amino) substituted *ortho*-carboranes are difficult to synthesize and isolate since, in the presence of amines, the *closo*-carboranes will decompose to their *nido*-cages. However, they can be stabilized as their ammonium halide salts or amide

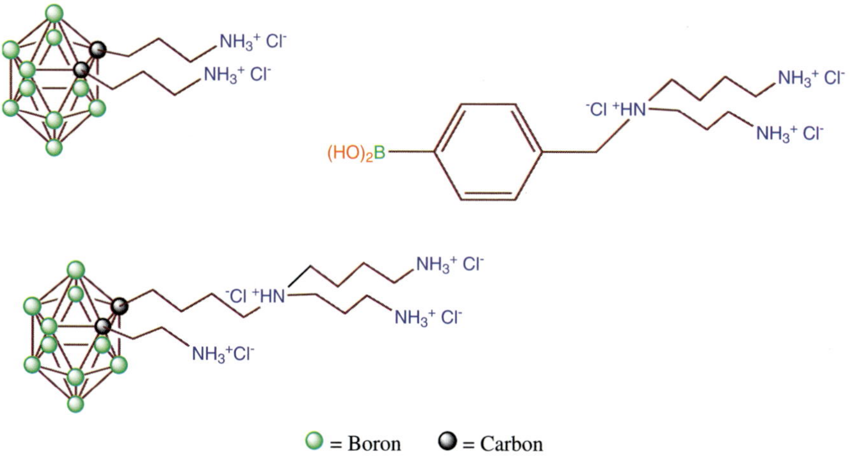

Figure 4.11   Boron-containing polyamines.

derivatives (see Fig. 4.11, for representative compounds).[120–122] The high toxicity of these amines and polyamines severely limit their use. To improve water solubility and create a potentially stronger interaction with DNA, the bis-substituted short-chain-amino-*ortho*-carboranes were synthesized, as shown in Scheme 4.3.[123] The advantages of these compounds lie in the almost unlimited variety for the attached short chains. They avoid the possibility of side decapitation reactions, nucleophilic addition reactions, and elimination reactions.[123,124] Their water solubility can be dramatically improved by conversion to the ammonium halide salts. The preliminary results of the *in vitro* killing effects of these compounds are promising. Compared with the currently used BNCT drug BSH, tumor cytotoxic effects have been approximately doubled under the same operating conditions. Moreover, the carborane cage in bis(short-chain-amino)-*ortho*-dicarbaborane hydrochlorides can be decapitated to give the water soluble *nido*-derivates.[123]

4.2.3. *Boronated amino acids and peptides as BNCT agents*

Since the boron-containing amino acid BPA has been extensively studied as a BNCT agent, a great deal of effort has been directed toward modifying

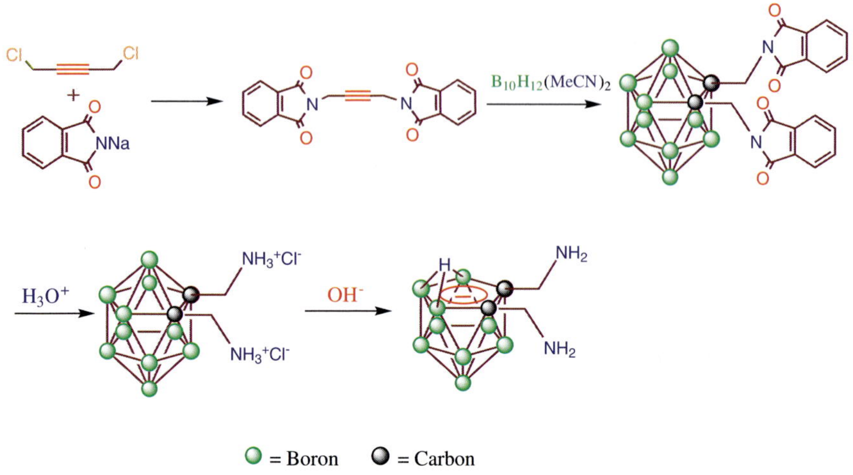

Scheme 4.3　Synthesis of short-chain carboranyl amine.

BPA (Scheme 4.1) and constructing peptides as well as proteins derived from these amino acids. The rational for expecting a higher concentration of the boronated amino acids in tumor cells compared to normal cells is that rapidly proliferating cells will require larger amounts such cellular building blocks.

Examples of reported modifications include: (a) increasing the BPA water solubility by the attachment of nonionic, hydrophilic groups, as seen in Fig. 4.12 (1).[125,126]; although a number of such compounds have been synthesized, little has been reported on their potential as BNCT agents; (b) replacing benzoid structures with high boron-containing clusters, as seen in Fig. 4.12 (3–6),[127,128] however, bioassay results demonstrated that carboranyl amino acids showed no evidence of tumor selectivity[130,131]; (c) constructing peptides from pristine or BPA analogues (see Fig. 4.12 (4)); this strategy was based on the projected increased need by tumor cells for protein precursors and the feasibility that small peptides may cross cellular membranes and be utilized by tumor cells; (d) constructing borane zwitterionic amino acid analogues such as H_3NBH_2COOH (Fig. 4.12 (2)), and their derived peptides, see examples in Fig. 4.12 (7,8).[131,133] Interest in such compounds primarily

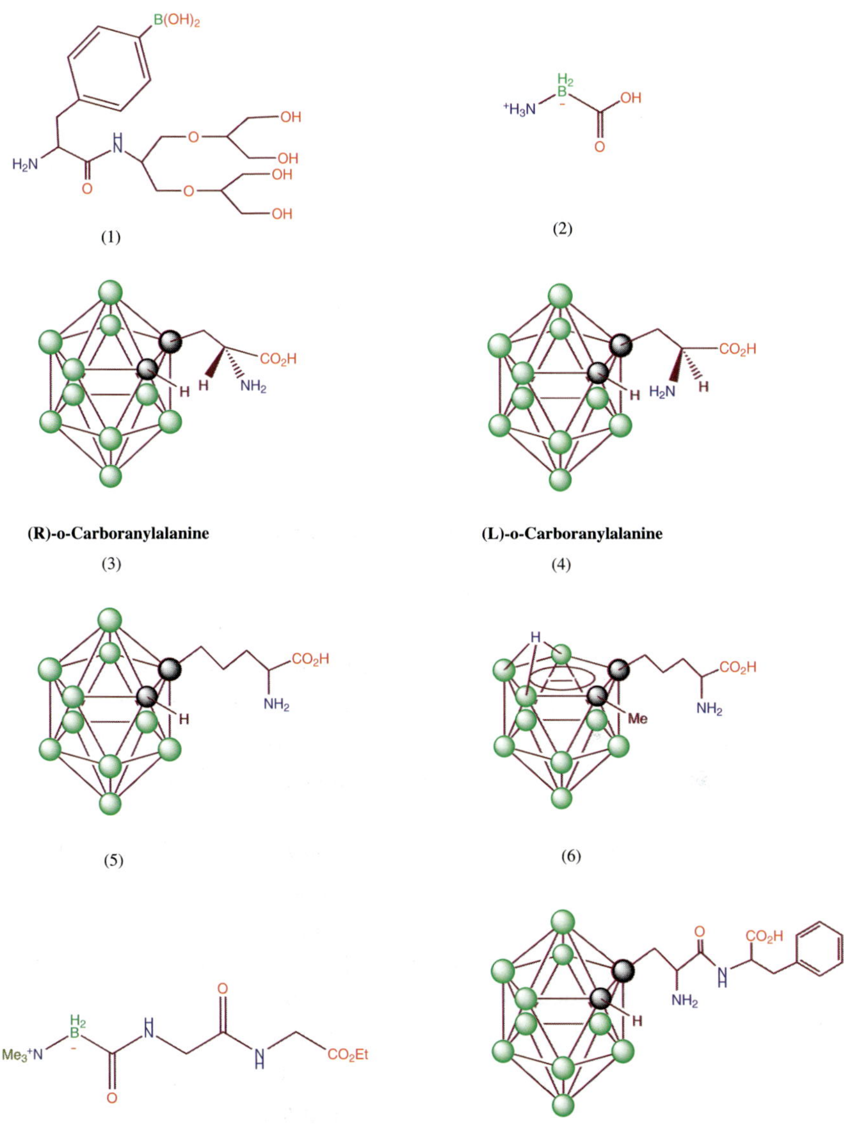

Figure 4.12 Boron-containing amino acids and derived peptides.

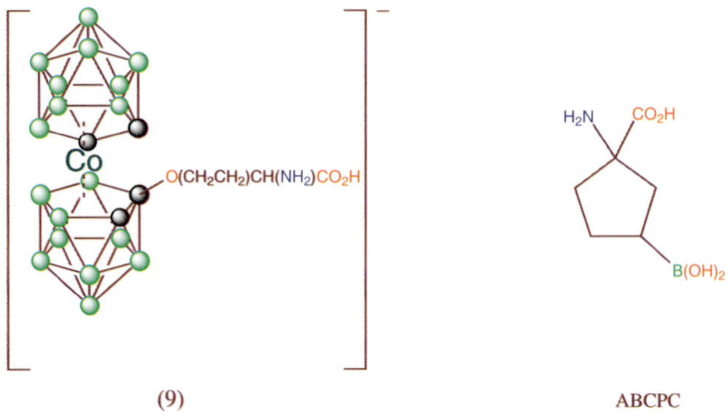

Figure 4.12   (Continued)

arose from their hypocholesterolemic and triglyceridemic activity and not solely from their BNCT potential. Research on the borane zwitterionic amino acids is still at its preliminary stage. On the other hand, some unnatural cyclic amino acids, such as 1-amino-3-borocyclopentanecarboxylic acid (ABCPC), has been found to be to be superior to BPA.[132b] In melanoma mouse model, ABCPC was found to have a tumor/blood ratio of 8 and a tumor/brain ratio of 21.

4.2.4. *Boronated nucleic acids as BNCT agents*

Nucleic acid is an essential cell metabolite precursor. The rationale for using boronated nucleic acids as BNCT agents is based on the increased requirements for DNA or nucleic acid precursors of rapidly dividing neoplastic cells.

Many approaches such as attaching boron species to purines, pyrimidines, and nucleosides, have been used to incorporate boron into the nucleic acid structures.[133–137] Some representative structures are shown in Fig. 4.13. Some nucleic acid-based BNCT agents, such as compound 3 in Fig. 4.13, were found to be able to penetrate the BBB and various cell membranes, and thus such boronated nucleosides were considered as potential BNCT agents for targeting brain tumors.[138] To improve the hydrophilicity of these agents, the carborane cages could be decapitated to form their anionically charged *nido* counterparts.[139,140] Alternatively, ionic functions could be inserted into the molecule or nonionic hydrophilic moieties can be attached.

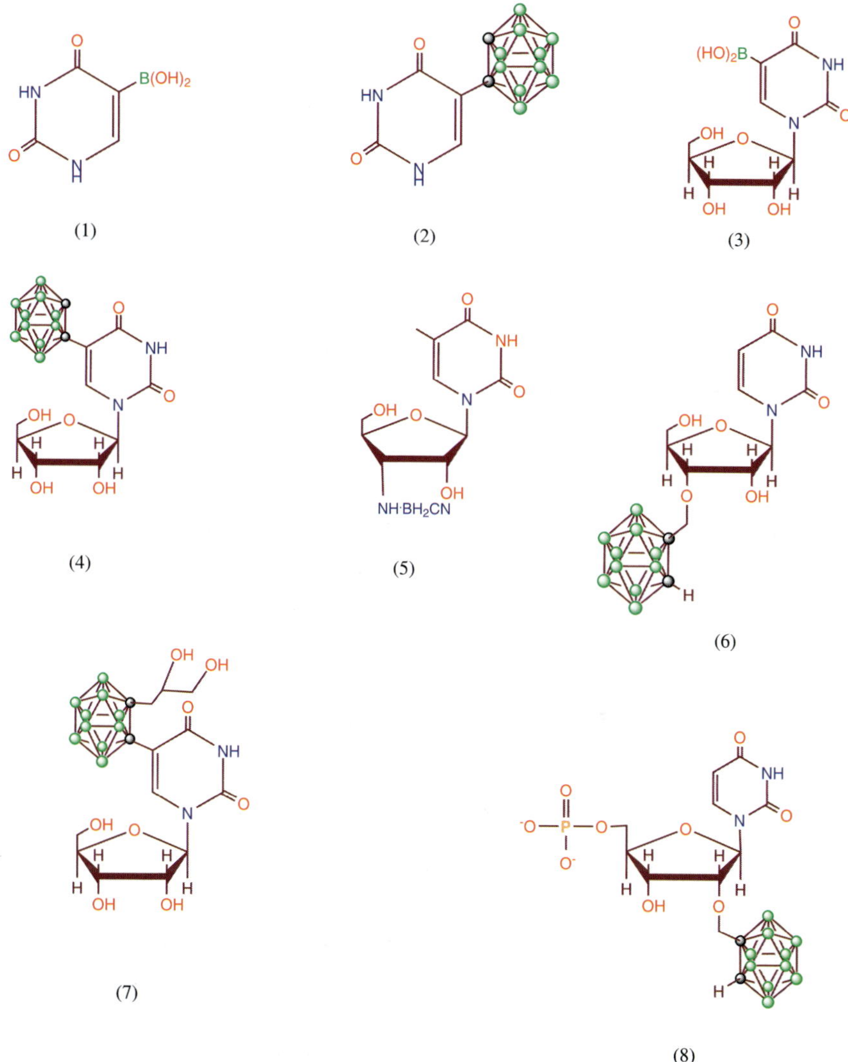

Figure 4.13   Representative structures of boronated nucleic acids as BNCT agents.

4.2.5. *Boronated carbohydrates as BNCT agents*

There are significant differences in the carbohydrate composition of the cell membrane surfaces of malignant cells when compared with normal cells.[141]

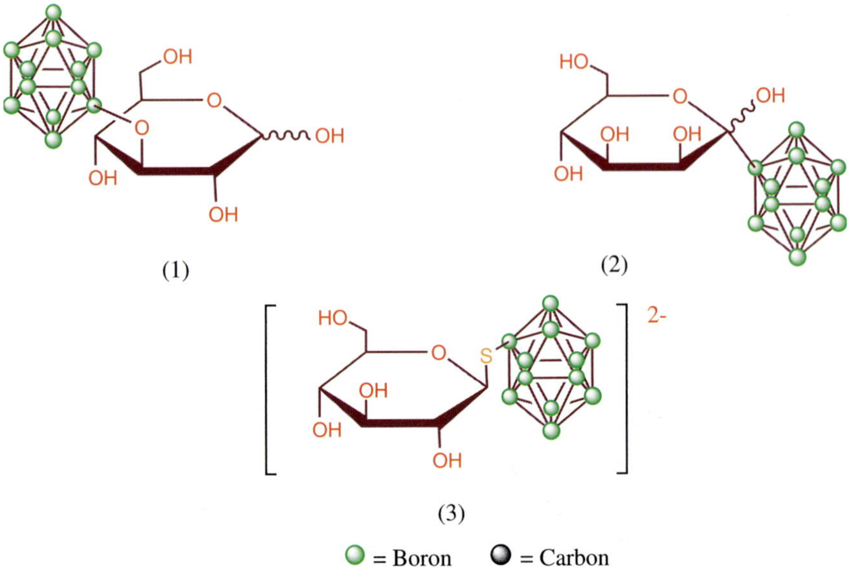

Figure 4.14 Carborane-containing analogues of carbohydrates.

This suggests the use of boron-containing carbohydrates to increase the concentrations of boron in tumor cells through the action of the glucose transport system.[142,143] In addition carbohydrate functionality of boron clusters can increase their aqueous solubility. Initially the combination of a boron cage and a carbohydrate moiety was developed for binding to antibodies.[144] Subsequently, various boron cages have been attached to carbohydrates (Fig. 4.14).[145,146] Unfortunately, there is as yet no evidence that these compounds behave as precursors of structural carbohydrates or that they use the active transport system to achieve differential uptake in tumors.[147] Their biodistribution evaluations are limited and much remains to be accomplished.

4.2.6. *Phosphorus-containing boron compounds as BNCT agents*

Phosphorus is an important component for all the animal tissues. It has been observed that various natural and synthetic ether lipids possess a proclivity for and persistence in a variety of tumors. This is partially due to

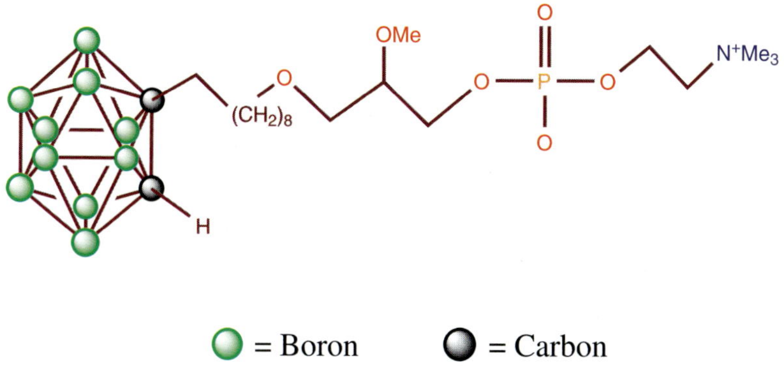

Figure 4.15   Carborane-containing phospholipid.

the absence of the enzyme O-alkylglycerol mono-oxidase in their neoplasms. Based on the high accumulation, it was anticipated that boron-containing lipids or phospholipids could be used as BNCT agents. Lemmen *et al.* synthesized the carborane-containing phospholipid shown in Fig. 4.15.[147] However, biotest results including both cell culture studies and in tumor-bearing animals were not encouraging.[147,148]

On the other hand, it has been found that ^{32}P-phosphate concentrated in, and persisted to an extraordinary high degree in, a variety of malignant tumors.[149] Significant accretion of the studied anion by various tumors with its apparent exclusion from normal brain, heightened expectations that promising boron-containing phosphorus compounds could be developed as suitable BNCT agents, and much effort has been expended on the syntheses of phosphorus-containing boron compounds. Many phosphate compounds, including neutral compounds, functionalized with boron species have been synthesized and evaluated.[150–153]

Figure 4.16 shows some representative structures. Bioassay results of these phosphates have shown a significant boron incorporation and persistence in the tumor when compared with levels observed in blood and normal brain at the same time intervals. Unfortunately, preliminary studies have shown that these compounds are highly toxic. More complete biological assessment of boron-containing phosphates and phosphonates as potential tumor-targeting BNCT agents still remains to be carried out.

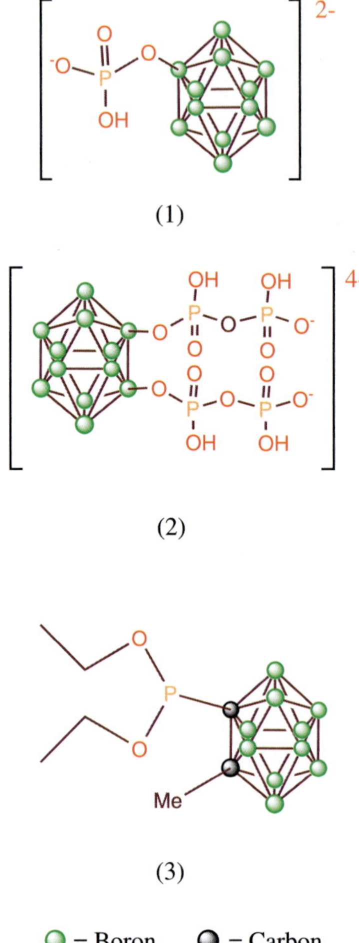

Figure 4.16   Carborane-containing phosphorus compounds.

4.2.7. *Other boron-containing compounds as BNCT agents*

In addition to those compounds discussed above, other compounds such as boron-containing porphyrins, phthalocyanines, and hormone analogues, have also been investigated as potentially utilizable BNCT agents. Since porphyrins were found to show both selective uptake and persistence within tumors, their boron-containing derivatives were synthesized

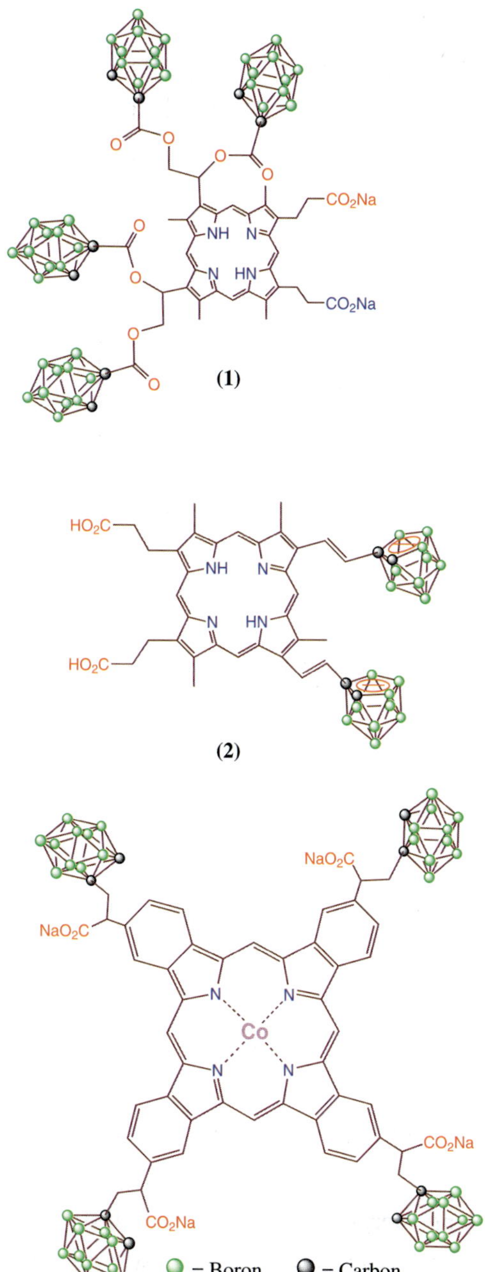

Figure 4.17 Boron-containing porphyrins.

with the expectation that they would be good BNCT vehicles. Some selected boron-containing porphyrins are shown in Fig. 4.17.[154–157] Both BOPP (Fig. 4.17(1)) and VCDP (Fig. 4.17(2)) were extensively studied in animals; however, *in vivo* studies demonstrated that they were toxic at the levels required to obtain useful tumor concentrations.[158(a)] Other porphyrin-based compounds have been developed with much lower toxicity.[158(b),159] A number of porphyrin-cobaltacarborane conjugates have been synthesized and their cell uptakes have been investigated.[160] The number and distribution of the cobaltacarboranes had an effect on the cellular uptake, in general the greater the number of metallacarborane groups on the porphyrin, the greater the cellular uptake.[160(a)] Cellular uptake could be enhanced by attaching cell penetrating peptide sequences to the porphyrin-cobaltacarborane conjugant.[160(b)]

The method of delivery was found to be important. Ozawa *et al.*[158b] reported that, for the porphyrin TABP-1 (see Fig. 4.18), when the maximum

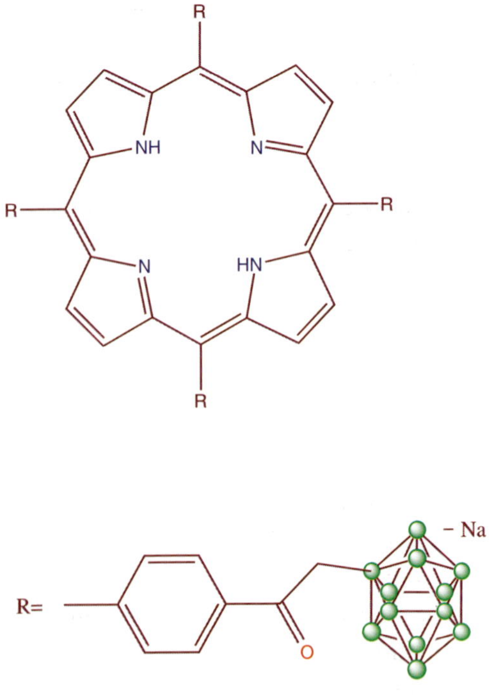

Figure 4.18 Chemical structure of TABP-1. Polyhedral cage is $B_{12}H_{11}$.

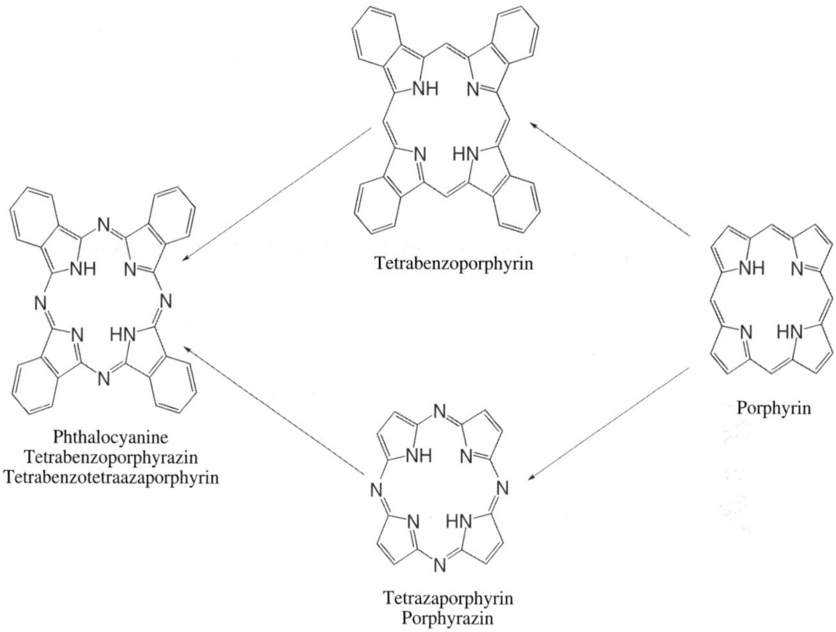

Tetrabenzoporphyrin

Phthalocyanine
Tetrabenzoporphyrazin
Tetrabenzotetraazaporphyrin

Porphyrin

Tetrazaporphyrin
Porphyrazin

Figure 4.19   Relationship between porphyrins and phthalocyanines.

tolerated dose of 15 mg/kg was injected into the tail veil of athymic rats bearing intracerebral human glibastoma xenographs, accumulation in tumor tissue was less than the 30 μg/g necessary for effective BNCT. However, the administration of 0.5 mg of TABP-1 by convection-enhanced delivery (CED) resulted in a tumor boron level of 65.4 μg/g of tumor with a 160:1 tumor to serum ratio. A number of different porphyins have been synthesized and studied as possible BNCT boron-delivery vehicles.[159]

Phthalocyanines, which are chemically similar to the porphyrins (see Fig. 4.19, for a comparison), have also been studied as the bases of boron carriers. Both the porphyrins and phthalocyanines have also been used in photodynamic therapy (PDT). PDT is another binary therapy in which photosensitive compounds are localized in cancer cells and then, in the presence of molecular oxygen, irradiated with red light to generate singlet oxygen, which is an active cell toxin. There have been studies aimed at

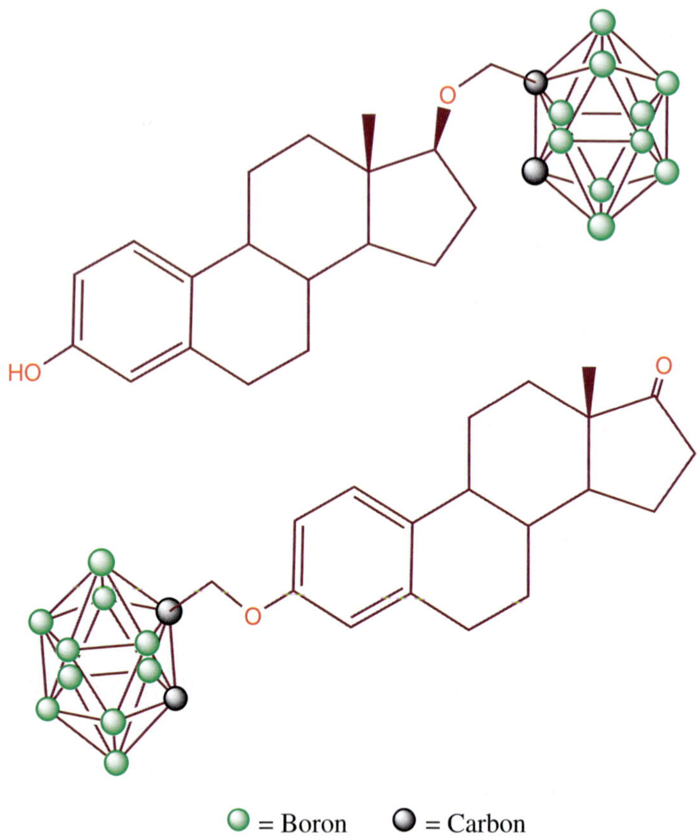

Figure 4.20   Boron-containing hormone and analogues.

developing compounds for treatment of cancer cells with a combination of BNCT and PDT.[161]

It is known that hormones are closely related with tumors in their development and proliferation. Some hormone receptors are localized in the cell's nucleus, so boron-containing hormone precursors could be used as BNCT agents. Representative structures of these compounds are presented in Fig. 4.20.[162–164] The advantage of hormone-derived BNCT agents lies in their relatively low molecular weights, which offers the potential for rapid tumor targeting. But, for such compounds to be effective, they must be nontoxic, stable under *in vivo* conditions, and exhibit sustained hormonal activities.

4.3. Boron-Containing Nanomaterial-Based BNCT Agents

Research on nanomaterials has grown explosively in the last few years, with an increased emphasis on developing nanomaterials as drug delivery vehicles.[165–167] The size (<1000 nm) of such delivery vehicles has attracted wide interest in the field of drug targeting. The high boron concentration needed for BNCT has generated a great deal of interest in nanomaterials, since, at least in theory, they could incorporate a large number of boron atoms, thereby lowering the dose requirement. For this reason, the syntheses and evaluations of boron-carrying nanomaterials as possible BNCT therapeutic drug delivery agents have received a great deal of interest.[168] Nanomaterial-based drug systems provide the advantages of being able to penetrate cell membranes through minuscule capillaries in the cell walls of the rapidly dividing tumor cells,[169] while at the same time having low cytotoxicity in normal cells. They have been found to have favorable interaction with the brain blood vessel endothelial cells of mice,[170] and are passed through the BBB. Their role as boron carriers can be further enhanced if the nanocarriers are capable of avoiding detection and clearance by the immune system, specifically the reticuloendothelial system (RES). The size of the nanocarriers, though offering advantages, also presents problems. For example, it has been shown that liposomes having diameters >50 nm have difficulty penetrating the BBB[171]; thus, the optimum dimension of the nanoparticles as drug carriers has become an important area of research. Currently, both the effectiveness and the application of several different types of nanoscale materials as drug targeting agents are under investigation.[168]

4.3.1. *Liposomes*

Liposomes are small, spherical vesicles composed of membranes of phospholipids. The phospholipids are molecules having a hydrophilic head and a hydrophobic tail. Cell membranes are composed of such molecules arranged in two layers. When these membranes are disrupted, they can reassemble as extremely small spheres, usually as bilayers (liposomes). Figure 4.21 shows a cutaway drawing of a bilayer liposome.[172] The membrane thickness is of the order of 4 nm and the vesicle can be composed of one bilayer (unilamellar)

Figure 4.21   Cutaway drawing of a liposome.[172]

or more bilayers (multilamellar). The interior aqueous volume can contain water-soluble drugs for delivery once the liposome enters a cell. In addition, hydrophobic compounds could be stored in the lipid layer[175]; liposomes have the potential of delivering large amounts of boron to cancer cells.[173,175]

One of the problems associated with the use of liposomes as drug delivery vehicles is the rapid rate of clearance by the reticuloendothelial system (RES). This decreases the circulation time of the liposomes and hence their localization in tumors. Studies have shown that the uptake of drugs encapsulated in liposomes is higher in tumor cells than in the surrounding normal cells. Prolonged circulation time can be achieved by modification of the bilayer composition of the liposome.[174] In addition, ligands, such as antibodies and their fragments, folate, transferrin and some peptides can be linked to the liposomes for specific tumor targeting.[176,177] BNCT agents can be encapsulated inside liposomes that circulate throughout the blood stream where the drug is then released via diffusion through the liposome or by liposomal degradation. Liposomes as potential BNCT delivery agents have been well investigated.[178] Hawthorne and coworkers have made major contributions in developing liposomes as boron-delivery agents to tumors.[179] Several of the compounds developed for this purpose are shown in Fig. 4.22.[175,179–181] For compound $Na_3[\alpha^2\text{-}B_{20}H_{17}NH_2CH_2CH_2NH_3]$

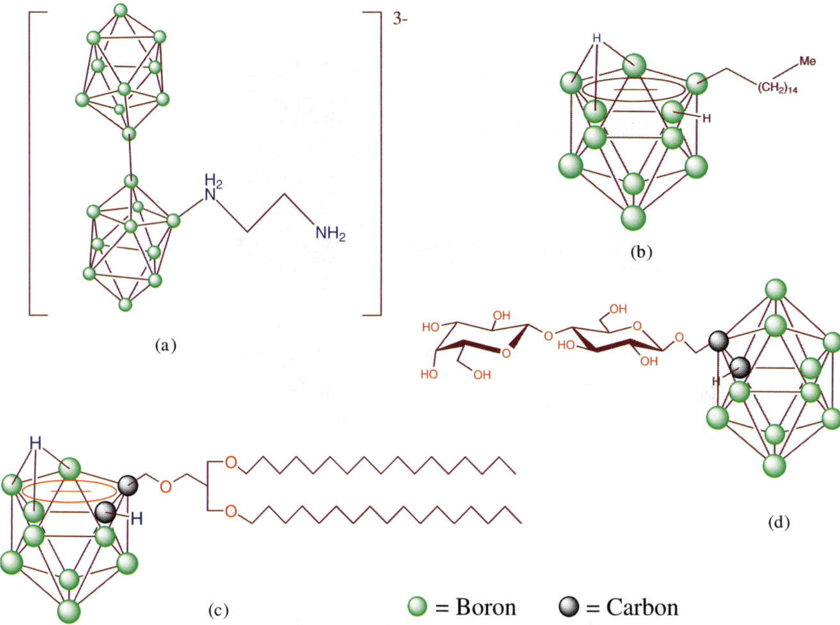

Figure 4.22 Structures of (a) amino polyhedral borane anion, (b) carboranyl anion, (c) dual-chain *nido*-carborane, and (d) sugar-derived carborane for liposome encapsulation.

(Fig. 4.22 (a)) as a BNCT agent, preliminary *in vivo* tests with mice demonstrated improved tumor uptake and retention.[180] Dual-chain *nido*-carborane compounds, such as the structure shown in (Fig. 4.22 (c)), have been used to prepare stable, high boron-containing liposomes.[175,181]

Cationic liposomes are currently used in gene therapy, because of their ability to target cell nuclei, and they have also been used as carriers for sugar-derived carborane compounds (Fig. 4.22 (d)) in BNCT. These new delivery systems may reduce the amount of boron required for successful BNCT treatment.[182]

It has also been demonstrated that the administration of ^{10}B-polyethylene-glycol (PEG)-binding liposomes, with mean diameters of about 116 nm, can suppress the growth of AsPC-1 tumors in nude mice implanted with human pancreatic carcinoma xenografts.[183] The conjugate was able to avoid the phagocytosis by the RES and, thus, could circulate in the body for longer periods of time. Additionally, the development of a new generation of magnetic liposomes is in progress. In this technology, liposomes are loaded with both a therapeutic drug and a ferromagnetic material. These magnetic liposomes could then act as drug carriers under the influence of an external magnetic field. Magnetic liposomes-containing doxorubicin have been shown to be rather effective drug carriers to osteosarcoma-bearing hamsters in that they increased the concentration of doxorubicin in the tumor.[184] However, the concentration of doxorubicin in the tumor appeared to have dropped several weeks after administration (see Fig. 4.23). In another study, magnetic particles were coupled to the chemotherapeutic agent mitoxantrone and directed to tumors implanted in the flanks of rabbits to give complete tumor remission. In addition to that, the applied dose of the drug could be diminished to 20% of the regular systemic dose.[185]

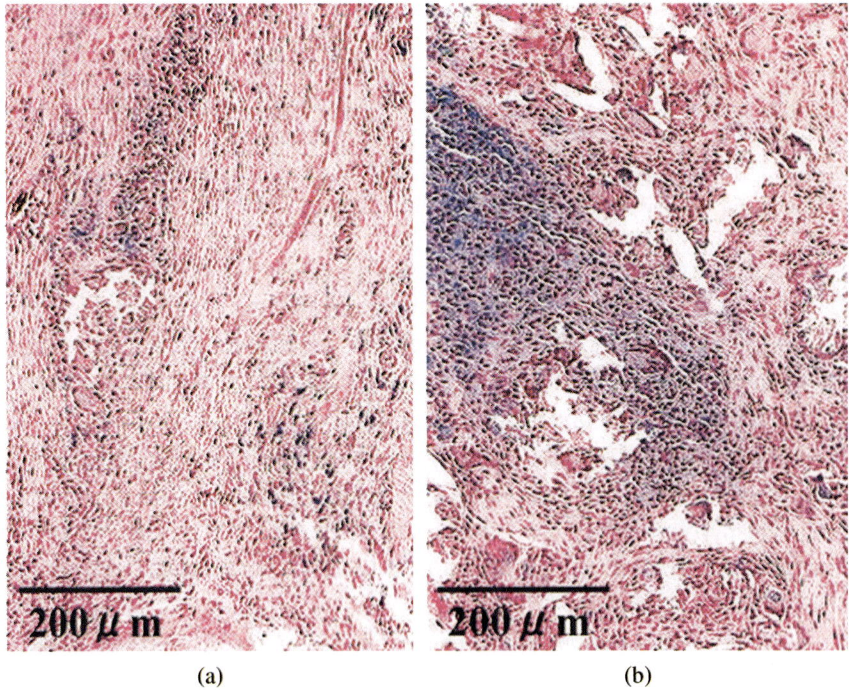

Figure 4.23   Light micrographs of Berlin blue-stained osteosarcoma excised 1 hr (a) and 2 weeks (b). After IV administration of magnetic doxorubicin liposomes under magnetic force of 0.4 T. Blue spots represent magnetite particles, and they are concentrated and deposited in the endothelium of proliferated tumor vessels and the tumor interstitium in (a) but not (b).[184]

4.3.2. *Dendrimers and dendritic polymers as BNCT agents*

The idea of using polymers to transport proteins was first raised in the 1970s; since then, polymer-derived drug delivery systems have attracted much attention.[186,187] Modified polymeric carriers can improve water solubility, prolong vascular circulation time, and therefore selectively increase the accumulation in tumor tissue via the enhanced permeation and retention (EPR) mechanism.[188,189]

Dendrimers are uniform-sized, globular macromolecules that have distinctive characteristics such as high degrees of branching, multivalency, unique molecular weight, and modifiable surface functionality.[190] They are built from a series of branches extending out from an inner core; each successive layer of branching is called a generation. They have the potential to carry therapeutic agents either in their internal cavities or on their periphery by means of covalent bonding.[191] Furthermore, they can be engineered to deliver either hydrophobic or hydrophilic molecules and thus offer a broad range of possibilities as therapeutic delivery agents.[192] A frequently used dendrimer scaffold is the polyamidoamine (PAMAM) dendrimer that is designed to have a water-soluble backbone and widely use in drug delivery. Figure. 4.24 shows a second generation PAMAM dendrimer.[193] The PAMAM dendrimers are biocompatible, nonimmunogenic, water soluble and possess modifiable surface amine groups for binding to various molecules. Anticancer drugs loaded onto PAMAM dendrimers conjugated to folic acid as a targeting agent were found to congregate in human KB tumors that overexpress folic acid receptors.[194]

Barth, *et al.* have reported the uptake of boronated epidermal growth factor (BSD-EGF), both alone and in conjunction with BPA, by epidermal growth factor receptor (EGFR)-positive glioma (F98$_{EGFR}$).[195] EGFR is amplified or mutated in many human gliomas but its expression is low in normal brain. The boron delivery agent (BSD) was a heavily boronated fourth generation starburst poly(amidoamino) dendrimer whose surface amino groups had been reacted with $Na(CH_3)_3NB_{10}H_9NCO$ (see Ref. 196). It was found that 24 hr after intratumoral (*i.t.*) injection of the BSD-EGF biocunjugate into Fischer rats bearing the F98$_{EFGR}$ gliomas, a tumor boron concentrations of 21.1 µg/g was found.[197] This is close to the 35 µg/g tissue deemed necessary for effective BNCT. BNCT studies

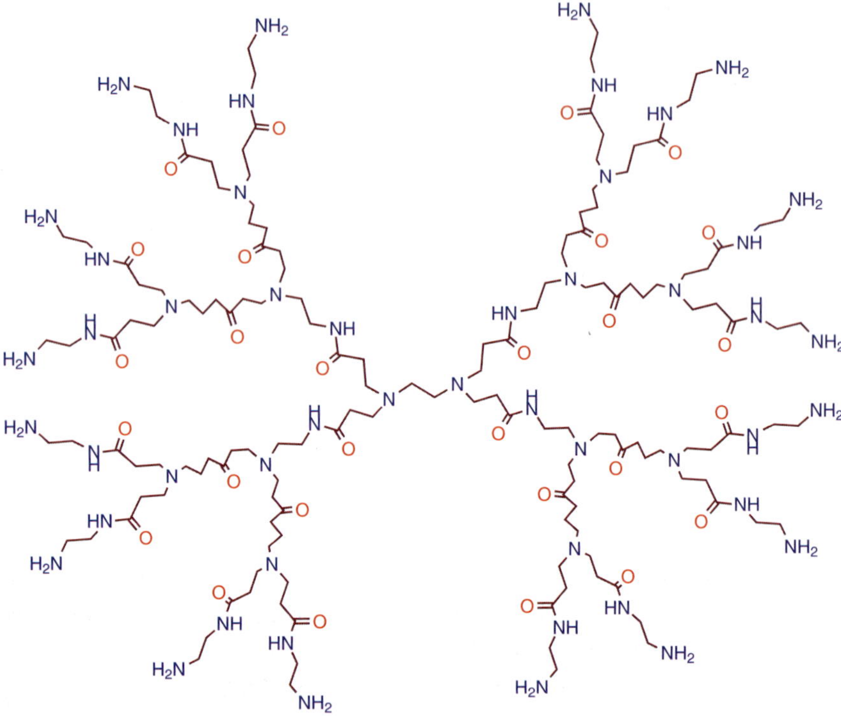

Figure 4.24   Second generation polyamidoamine (PAMAM) dendrimer.

showed that the mean survival time (MST) of animals having $F98_{EGFR}$ gliomas was 45 ± 5 da compared to 33 ± 2 da for animals having $F98_{WT}$ gliomas, which were EGFR negative. On the other hand, simultaneous BPA and BSD-EGF injection resulted in a MST of 57 ± 8 da, which was not statistically different from that of just BSD-EGF.[197]

In other studies, the chimeric MoAB cetuximab (IMC-C225), which is directed against (EGFR)-positive gliomas, were used to treat $F98_{EGFR}$ and a mutant form of F98 ($F98_{EGFR-VIII}$) tumors (see Fig. 4.25).[198] A heavily boronated generation 5 PAMAM dendrimer ($G\text{-}5B_{1100}$) was linked to the IMC-C255 to give the bioconjugate, $C225\text{-}G\text{-}5B_{1100}$ (BD-C225) which was tested on rats with $F98_{EGFR}$ and $F98_{WT}$ intracerebral gliomas implants. Results showed that 24 hr after direct intratumoral injection,

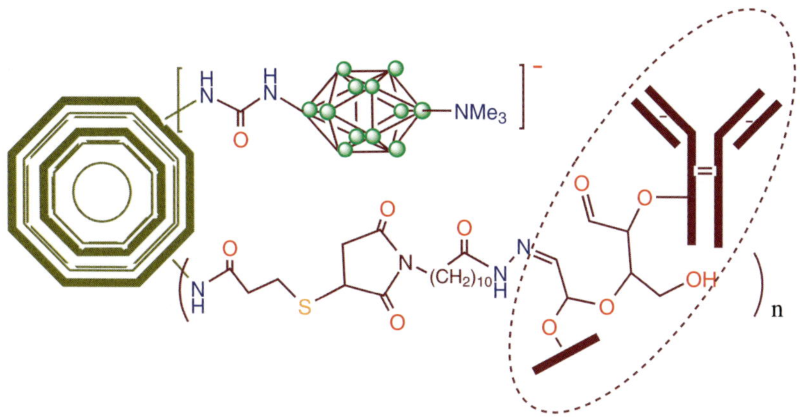

Figure 4.25 PAMAM G-5 dendrimer coupled to a IMC-C225 MoAB.

92.3 ± 23.3 µg B/g tumor was localized in F98$_{EGFR}$ tumor, 36.5 ± 18.8 µg B/g tumor in F98$_{WT}$, and 13.4 ± 6.1 µg B/g in normal brain.[143] EGFR-VIII is expressed and amplified in up to 54% of human glioblastomas and 75% of anaplastic astrocytomas.[199] Therefore, several mAbs have been directed against this receptor, one of which is L8A4, as described by Bigner and coworkers.[200]

A heavily boronated fourth or fifth generation PAMAM dendrimer was covalently attached to the L8A4 mAb and administered by CED into the brains of Fischer rats. which were previously IC implanted with receptor negative F98$_{WT}$ gliomas or receptor positive F98$_{npEGFR-VIII}$ giloma cells.[201] The results of BNCT studies on F98$_{npEGFR-VIII}$ giloma cells showed that rats receiving the BD-L8A4 bioconjugate and IV BPA had a MST of 85.5 ± 15.5 days with 20% long term (>6 mo.) survivors compared to a MST of 70.4 ± 11.1 days with 10% long-term survivors for those receiving the BD-L8A4 bioconjugant alone, 40.1 ± 2.2 days for those receiving only IV BPA, and 26.3 ± 1.1 days for the controls.[201] A combination of both C225-G-5B$_{1100}$ and BD-L8A4 delivered by CED to rats infected with F98$_{WT}$ and F98$_{npEGFR-VIII}$ giloma cells, followed by BNCT had a MST of 55 days, compared with 36 days for BD-L8A4 and 38 days for BD-C225 alone.[202] Since it is highly unlikely that all tumor cells in glioblastoma multiform would express either EGFR-VIII and/or

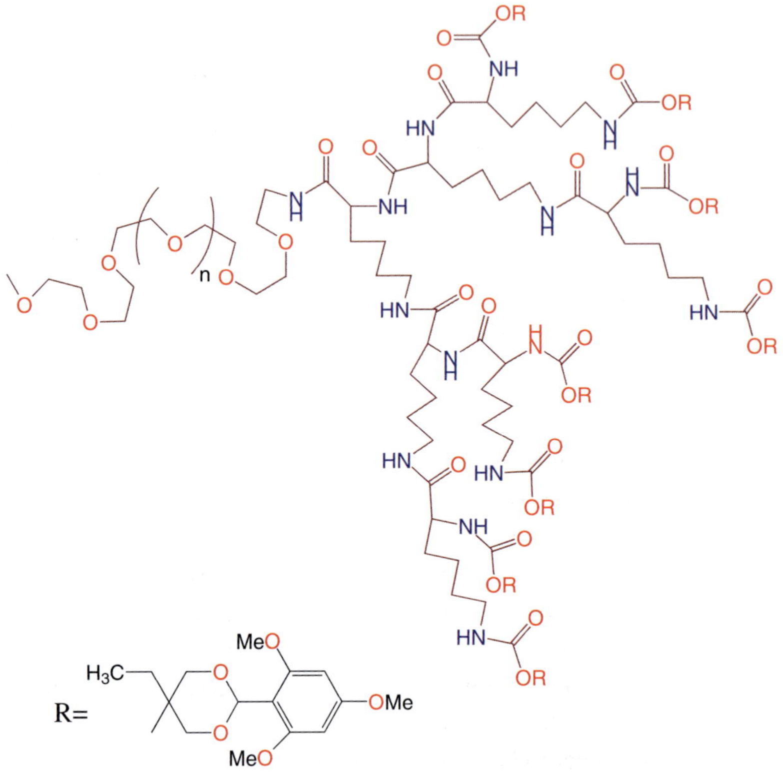

Figure 4.26 pH-sensitive bioconjugant molecules.

EGFR, it was recommended that combinations of the two conjugates plus BPA or BSH be administered.[203]

In addition to covalently attaching drug molecules to the surfaces of dendrimers, it is also possible to entrap drug molecules in the interiors of dendrimeric substances. However, problems arise with dendrimers that encapsulate drugs in their core. Such entrapped drugs present problems associated with leakage and the difficulties encountered during the drug release process which is not easy to control. There have been recent reports on the development of pH responsive dendrimeric copolymers (see Fig. 4.26).[204]

These involve the construction of polyethylene oxide (PEO) hybrids and dendrimers tagged with pH-sensitive hydrophobic acetal groups on the periphery. Upon hydrolysis at slightly acid pHs, the hydrophobic groups

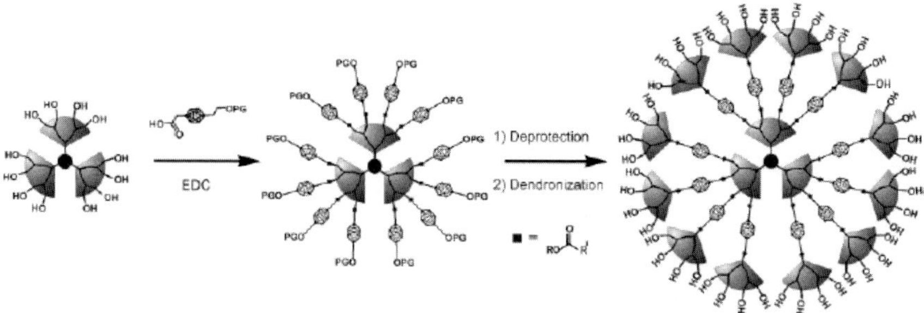

Figure 4.27   Aliphatic polyester dendrimer-containing 12 interior carborane cages.[206]

were lost, which prompted the release of the load. Figure 4.26 shows two such examples, PEO-5K-[G-3]-polylysine-acetyl (5a) and PEO-10K-[G-3]-polylysine-acetyl (5b). Both of these constructs were stable at pH~7 but the acetyl groups hydrolyzed readily at pH 5.[204] Similar PEO-dendritic polyester hybrids were loaded with the anticancer drug doxorubicin (DOX) and found to exhibit similar pH sensitivity. Results with BALB/c mice bearing s.c. C-26 tumors showed essentially complete tumor regression and 100% survival of the mice over a 60 days period.[205] This novel drug delivery system has yet to be applied to BNCT, but deserves study.

There have been reports of the incorporation of carborane cages into the interior of dendrimeric structures. Figure 4.27 shows an example of a dendrimer containing 12 *p*-carboranes incorporated into a polyester dendremic core, this was used to prepare fourth and fifth generation dendrimers containing 4, 8, and 16 carboranes in their interior, with hydroxyl groups on their periphery. Water solubility was assured as long as there were eight hydroxyl groups per carborane.[206] Interiorly bound carborane dendrimers can also be produced by reacting tetra- and dodeca-alkyne unimolecular micelles (micellanes) with decaborane ($B_{10}H_{14}$) to give the corresponding tetra- and dodeca-*o*-carboranes. Water solubility could be imparted by converting the peripheral hydroxy-groups to their corresponding sulfates.[207]

It has been pointed out that all the boron delivery agents, such as liposomes and dendrimers, are inefficient, in that boron only constitutes a small percentage of the molecules. Hawthorne, and coworkers, have described the construction of heavily boronated closomer delivery agents

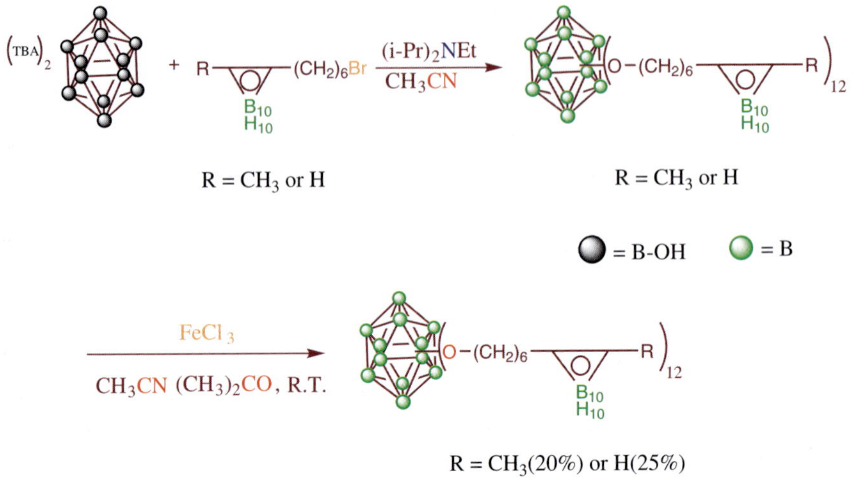

Figure 4.28 Synthesis and oxidation of dodeca[6-(2-R-1,2-dicarba-*closo*-dodecaboran-1-yl)hexyl]-*closo*-dodecacarborane(2-) (R = Me, H) closomers.

starting with the perhydroxolated $[closo\text{-}B_{12}(OH)_{12}]^{2-}$ and attaching carborane cages through either ether or carboxyl linkages. Figure 4.28 summarizes the synthesis of the closomer constructed through ether linkages. The carboranes could be deboronated by NaCN to give the corresponding *nido*-carboranes having a total — 14 charge.[208] Animal studies of the *nido*-closomer, dodeca[6-(7,8-dicarba-*nido*-dodecaboran-7-yl]-*closo*-dodecaborane(14-), showed it to be nontoxic at a therapeutic dose of 10 mg B/kg body weight. Murine biodistribution studies using BALB/c mice bearing EMT6 tumors showed that the boron uptake by the tumor reached a maximum of 10.5 μg B/g tumor 24 hr post-injection, with a tumor to blood ratio of 9.4. With the exception of liver and kidney other organs showed minimal boron uptake.[208] These encouraging results indicate that such *nido*-closomers are promising BNCT agents.

4.3.3. *Boronated carbon nanotubes as BNCT agents*

Carbon nanotubes (CNTs) are allotropic forms of carbon in which graphene sheets of sp^2 hybridized carbon atoms are rolled into a cylinder. Carbon nanotubes are either single wall (SWCNT), having a single graphene sheet, or multiwall (MWCNT) composed of a number of concentric cylinders. Since their discovery in 1991,[209] CNTs have attracted wide attention, with numerous applications proposed for their use.[210] Functionalized *f*-CNTs possess the highly favorable property of being able to enter various cells without showing apparent toxic effects.[211] For instance, it has been reported that peptide-functionalized single-walled carbon nanotubes were able to cross cell membranes and concentrate in the cytoplasm of 3T6, 3T3 fibroblasts, and phagocytic cells, without causing their death or inflicting other damages.[212] Similar results were obtained in HL60 cells, where it was found that functionalized SWCNTs could help transport large attached groups into cells without exhibiting cell toxicity.[213] *In vitro* studies of folic acid and fluorescent tag-conjugated SWCNTs have shown them to be effective in targeting HeLa cells, which are rich in folic acid receptors and are thought to stem from cervical cancer.[214] The *f*-CNT has the ability to passively cross cell membranes.[211] The reports of enhanced water solubility of CNTs through sidewall derivation with biologically important moieties[215–227] led to the exploration of the feasibility of using suitably derived SWCNTs as boron delivery agents for use in BNCT.[228] *Nido*-carborane units were attached to the sidewalls of SWCNTs to produce high boron content, water soluble SWCNTs (Fig. 4.29).[228]

These were then used to treat mice bearing the EMT6 tumor cells, a mammary carcinoma. A favorable tumor-to-blood ratio of 3.12 and a boron concentration of 21.5 µg B/g tumor were obtained after 48 hr of administration. In addition, it was observed that retention in the tumor tissue was higher than in the blood and other tissues including lung, liver, and spleen. The initial results are sufficiently good as to warrant further study.

In addition to sidewall attachments, it is possible to encapsulate boron carriers in the empty cavities of CNTs.[229] It has been reported that *ortho*-carboranes were successfully encapsulated inside SWCNTs (see below Fig. 4.30).[229] The composites were produced by mixing SWCNT and *o*-carborane in an ampoule and heating the 350°C in a tube furnace. The filling yields were quite low, varying from 5 to 20%. If high yield methods could be developed for producing this material, it could develop into an effective carrier for BNCT treatment.

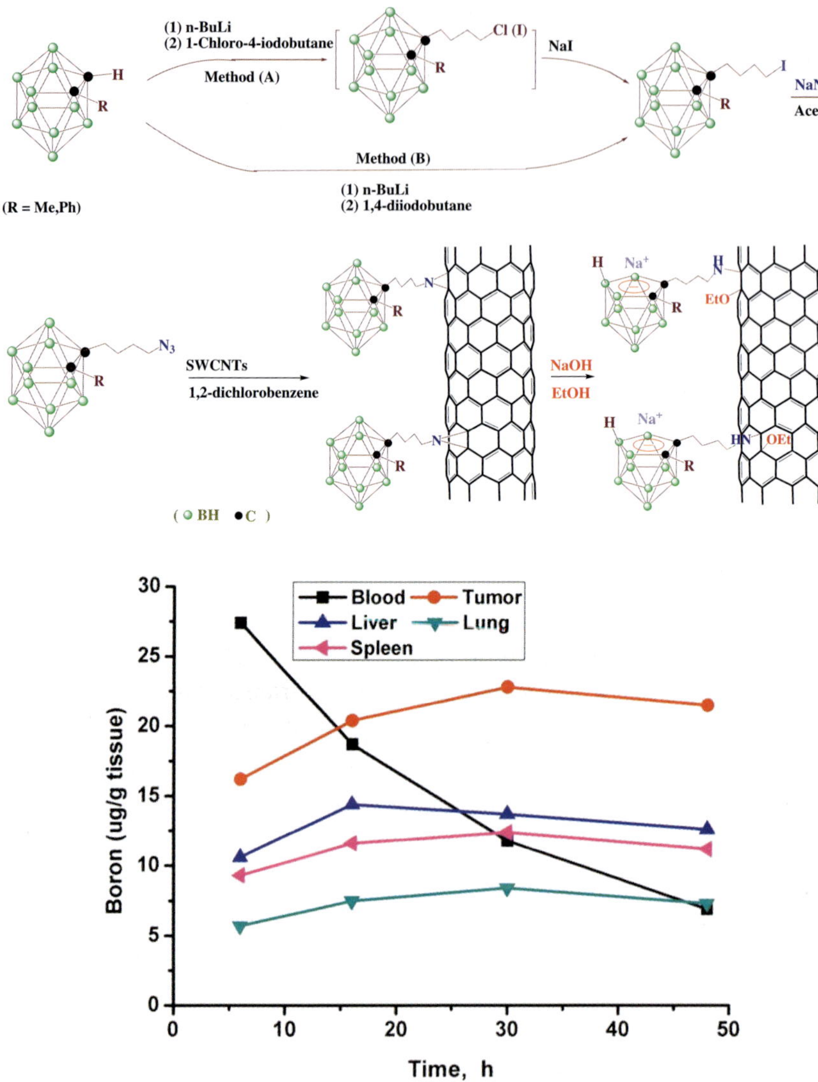

Figure 4.29 Synthesis and distributions of boron by SWCNTs-supported *nido*-carboranes in saline in different tissues.

Overall, the use of carbon nanotubes in future generations of BNCT drug delivery systems may improve therapeutic effectiveness and decrease undesirable side effects. Nevertheless, the long-term toxicity of

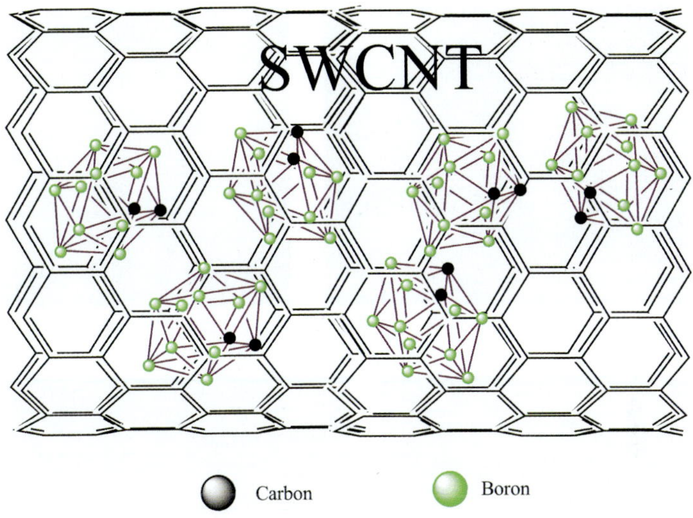

Figure 4.30 Schematic representation of *o*-carborane molecules within a SWCNT.[229]

carbon nanotubes must be thoroughly assessed. Further research is also expected to reveal the detailed mechanisms for transport across cell membranes. Other issues to be addressed include devising inexpensive, large-scale methods and techniques to embed small drug molecules, proteins, peptides, and genes inside and onto carbon nanotubes.

4.3.4. *Boron nanotubes as BNCT agents*

Another possible nanomaterial for use as a BNCT agent is the boron nanotube (BNT). BNTs are one of the latest nanomaterials resulting from advances in nanotechnology. Interest in their fabrication is largely due to the variable conductivity limitations present in the better-documented carbon nanotubes (CNTs). The BNTs can lead to the creation of a series of boron nanostructures, such as boron nanoribbons[230] and nanowires[231] (see Fig. 4.31).

Ciuparu *et al.* reported the first successful synthesis of a single-wall BNT in 2004.[232] In this method, pure single-walled BNTs were produced by the reduction of BCl_3 with hydrogen over a Mg-MCM-41 mesoporous catalyst. The diameters of the nanotubes were 3 nm, approximately the

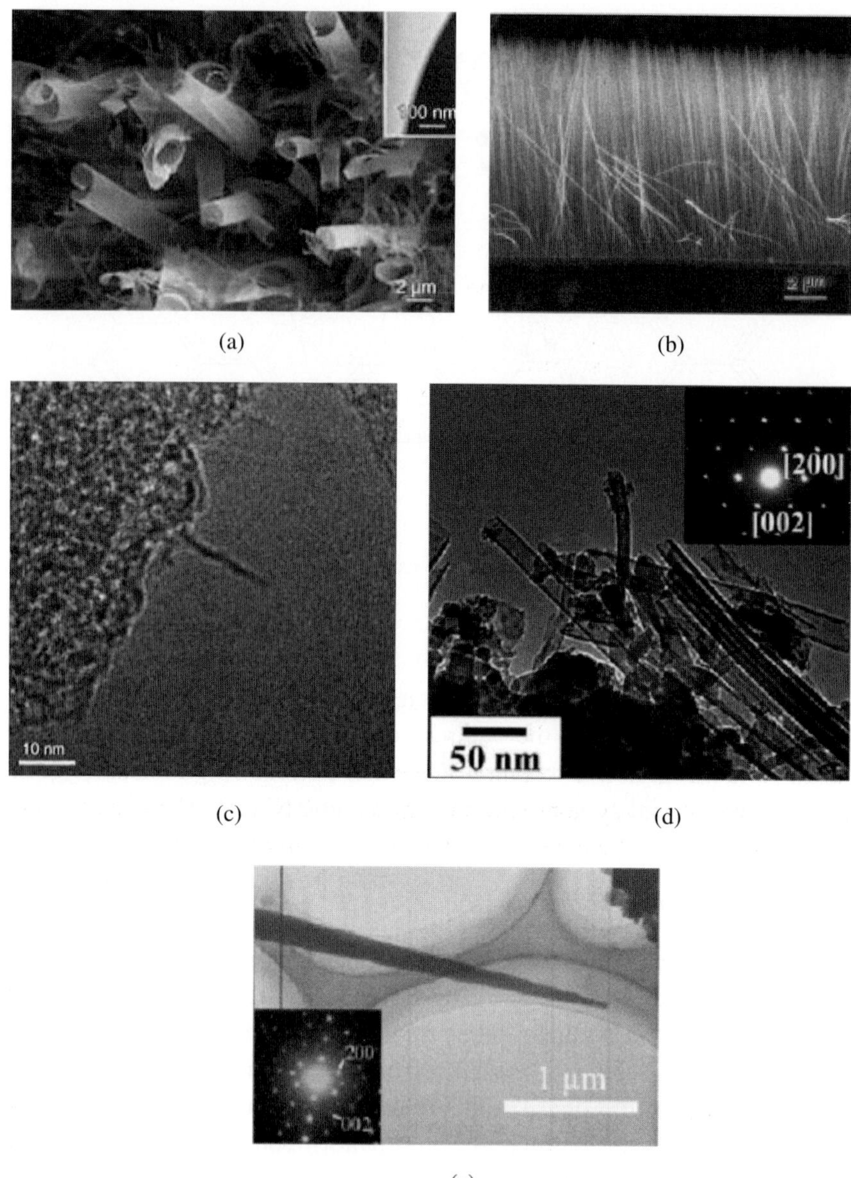

Figure 4.31   Micrographs of boron nanomaterials: (a) SEM image of boron nanorib-bons;[230] (b) SEM image of boron nanowires[234]; (c) TEM image of BNTs[232]; (d) SEM image of BNTs[233c]; (e) SEM of boron nanocones.[233b]

pore diameter of the mesoporous catalyst. A more general method of synthesizing low dimension boron nanostructures in large quantities from B/B_2O_3 mixtures has been reported by Gao and coworkers.[233] Under different conditions crystalline boron nanowires,[233a] nanocones,[233b] and nanotubes[233c] were produced. The nanowires were grown on a Fe_3O_4-coated Si(001) substrate from a 2:4:1 mixture of B_2O_3, B, graphite, respectively. The mixture was slowly heated under an Ar(g) flow to a temperature of 1000–1100°C and the nanowires grew on the substrate via a vapor–liquid–solid (VLS) process.[233a] The nanocones were grown on an Fe_3O_4-coated Si(111) substrate from a solid mixture of pure B_2O_3 and B in a 1:5 molar ratio at 1000–1100°C by the passage of a mixture of 95% Ar and 5% H_2.[233b] On the other hand, single crystal BNTs were obtained over a substrate coated with Fe_3O_4 catalyst from heating a B_2O_3 and B mixture in a 4:3 mass ratio, respectively, at 1000–1200°C under a flow of Ar gas.[233c] Figure 4.31 shows the SEM micrographs of some of these structures. Suitably functionalized, all are potential high boron content delivery vehicles.

Boron nitride nanotubes (BNNTs) are isosteres of CNTs and thus can duplicate some of their properties, such as hydrolytic stability and ease of surface functionalization; they also have high boron content (>40%) and could be used as boron delivery agents. In addition, BNNTs have been found to be inherently nontoxic, in contrast to CNTs.[235] They can be surface functionalized by the noncovalent adsorption of bioactive molecules. For example, BNNT-coated with second-generation glyco-dendrimers with α-mannose (G-2-man) was found to be stable for weeks in aqueous solution.[235] BNNTs, solubilized by coating with polyethyleneimine, showed good cytocompatibility with human neuroblastoma cells; cellular uptake was found to be by energy dependent endocytosis.[236] Further work by the same investigators showed that BNNTs could also be solubilized by coating with the protein poly-(L-lysine) (PLL). The PLL-BNNT was functionalized with a fluorescent molecule to allow tracking and with folic acid as a targeting ligand against tumor cells (see Fig. 4.32).[237] *In vitro* studies showed that the folic acid functionalized PLL-BNNT was selectively absorbed by glioblastoma multiforme cells in the presence of normal human fibroblast cells.[237] If such a preference is found *in vivo*, then the BNNTs could prove to be highly efficient BNCT agents.

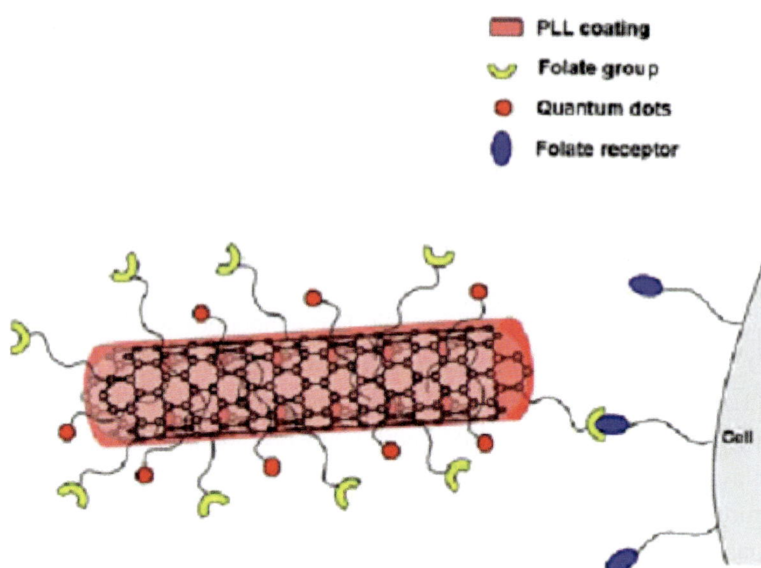

Figure 4.32   Proposed operation of the BNNT vector.[237]

4.3.5. *Boronated magnetic nanoparticles as BNCT agents*

A major obstacle in chemotherapy and drug discovery is that most of the therapeutic compounds are not very tumor-specific. When administered intravenously, they will generally distribute in and attack both carcinogenic and healthy cells, which results in deleterious side effects. In magnetically targeted therapy (see Fig. 4.33), a drug is attached to a biocompatible magnetic nanoparticle carrier, usually in the form of a ferrofluid, and is injected into the patient via the circulatory system.

When these bioconjugates enter the blood stream, external, high-gradient magnetic fields can be used to concentrate them at a specific target site within the body. Once the drug/carrier is correctly concentrated, the drug can be released, either via enzymatic activity or by changes in physiological conditions, and be taken up by the tumor cells.[238] The advantage of this methodology is that the required amount of the cytotoxic drugs is greatly decreased, thus reducing the associated side effects. Phase 1 clinical trials showed the method to be well tolerated in human subjects.[239] This

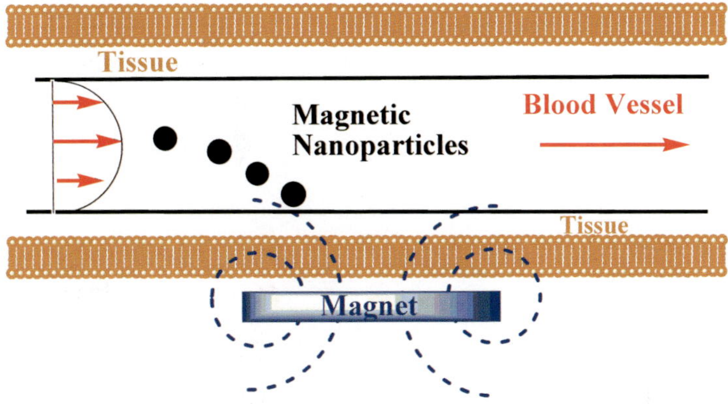

Figure 4.33 Magnetic drug delivery mechanism.

technique has been examined as a means of delivering cytotoxic drugs to brain tumors. Studies have demonstrated that particles as large as 1–2 µm could be concentrated at the site of intracerebral rat glima-2 (RG-2) tumors; a later study demonstrated that 10–20 nm magnetic particles were even more effective in targeting these tumors in rats.[240] Studies of magnetic targeting in humans demonstrated that the infusion of ferrofluids was well tolerated in most of the patients, and the ferrofluid could be successfully directed to advanced sarcomas without associated organ toxicity.[239] In addition, the preclinical and experimental results indicated that it is possible to overcome the limitations associated with magnetically targeted drug delivery, such as the possibility of embolization of the blood vessels in the target region due to accumulation of the magnetic carriers.[241]

Recently, FeRx Inc. was granted fast-track status to proceed with multicenter Phases I and II clinical trials of their magnetic targeting system for liver tumors. Therefore, applications of this technique can be considered as appropriate vectors for the use of BNCT treatment. BNNTs containing Fe catalysts (see Fig. 4.34) have been synthesized and their magnetic properties have been studied.[242a] *In vitro* studies have demonstrated the feasibility for influencing the uptake of the BNNTs by human neuroblastoma SH-SY5Y cells through the use of external magnetic fields.[242a] Therefore, the magnetic BNNTs have the potential for use as nanovectors for magnetic drug targeting, including usage in BNCT.

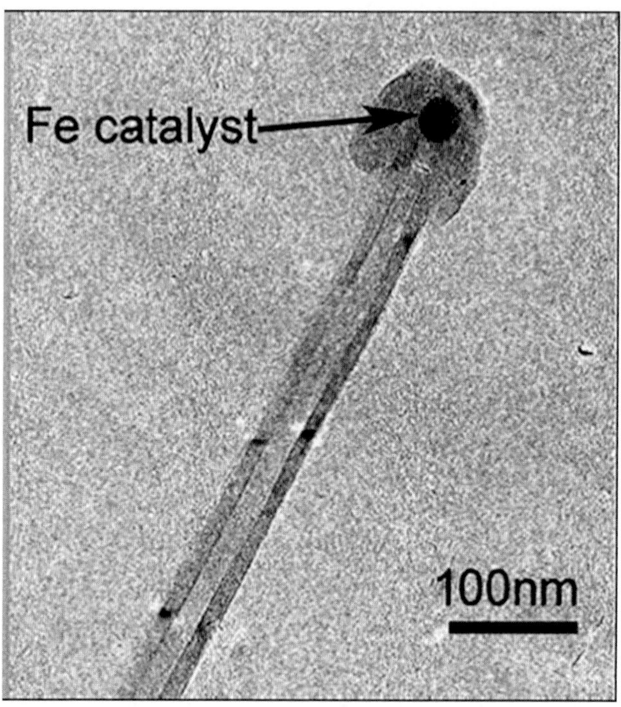

Figure 4.34   TEM image of a multiwalled BNNT with an Fe-particle at one tip.[242a]

An alternative method of magnetizing BNNTs was described by Huang *et al.*, in which an ethanol-thermal process was used for *in situ* formation of dense and uniformly distributed Fe_3O_4 nanoparticles on the surfaces of multiwalled BNNTs.[242c] By virtue of the magnetic Fe_3O_4 coatings, the BNNTs could be physically manipulated in a relatively low magnetic field.

In another study of magnetically directed nanoparticles for BNCT, *nido*-carborane cages were successfully attached to polymer functionalized ferrofluids and used in *in vivo* testing.[243] The bioconjugate was prepared by first attaching *closo*-carborane cages to propargyl-group-enriched magnetic nanoparticles through the catalytic azide-alkyne cycloaddition between 1-R-2-butyl-*ortho*-$C_2B_{10}H_{10}$ (R = Me, Ph) and a starch-stabilized magnetic iron oxide nanoparticle, as outlined in Scheme 4.4.[243] A loading amount of 9.83 mmol boron atom/g starch-matrixed magnetic nanoparticles has been reached. The resulting nanocomposites were found to

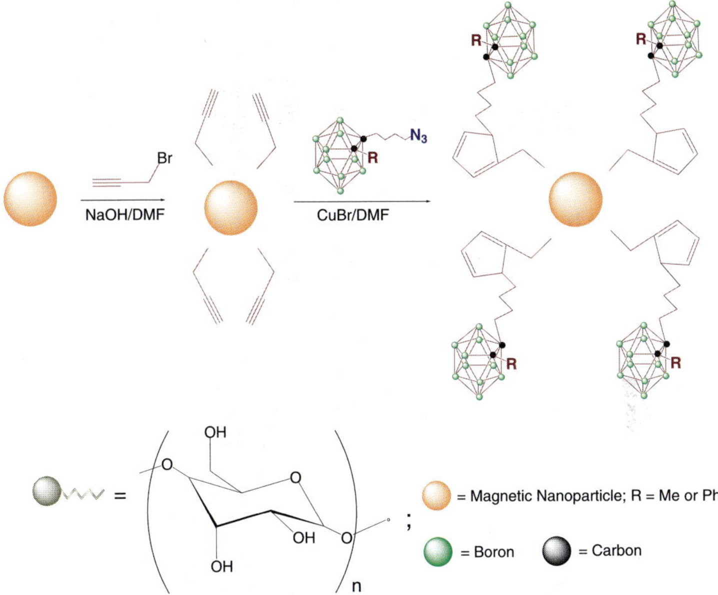

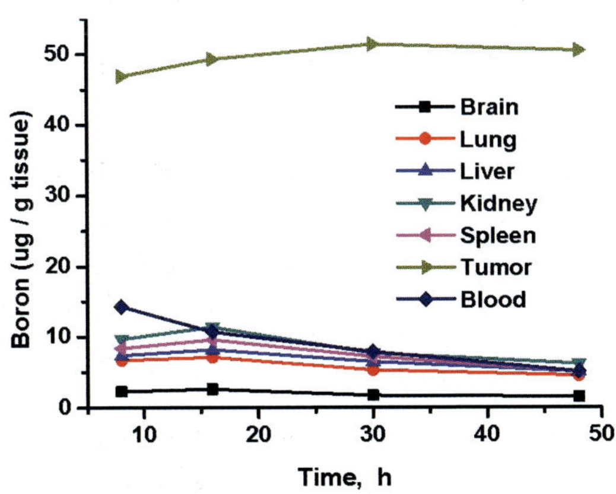

Scheme 4.4   Synthesis of encapsulated magnetic nanocomposites.[243]

Figure 4.35   Boron tissue distribution of encapsulated magnetic nanocomposite with external magnet.[243]

be efficient tumor-targeting vehicles under the influence of external magnetic field (1.14T), resulting in a high boron concentration of 51.4 µg/g tumor and high ratios of tumor to normal tissues concentrations (~10:1) (see Fig. 4.35).

Although the initial results are quite encouraging, issues related to magnetic nanoparticles, such as particle size control, surface fictionalizations, and their environmental compatibility need to be addressed.

4.4. Major Gadolinium Neutron Capture Therapy Drug Types

4.4.1. *GdNCT agents*

As pointed out earlier, GdNCT is also a promising approach to treat cancer. As with BNCT, one of the essential factors for the success of GdNCT is to deliver a sufficient amount of Gd selectively inside tumors for a fixed period of time. These factors are the same as encountered in BNCT and, as will be seen, many of the approaches discussed in BNCT will be encountered again in GdNCT. There are two extra complications: (1) in order for the Auger and internal conversion electrons to be therapeutic (cytotoxic), their extremely small LET's require placement close to the DNA; and (2) Gd^{3+} itself is toxic and it must be present within a complex.

For successful GdNCT it is necessary to deliver and maintain a sufficient amount of Gd within the tumors; this has proved to be a formidable task. Although hydrophilic agents, such as megluminegadopentate (GdDTPA, see Table 4.1), accumulate in tumor and are used as MRI enhancing agents; they are not suitable for GdNCT because of their rapid wash-out from the tumor.

The expanse of the malignant tumor cells is difficult to define due to their high invasiveness. Generally, one can define both a gross tumor volume (GTV) and a clinical tumor volume (CTV). The GTV defines the macroscopic tumor extent that is demonstrated by CT or MRI scans; the T1-enhanced MRI scan (shown to the left in Fig. 4.36) delineates the GTV. However, the tumor cells have already invaded the peri-tumor lesion surrounding the tumor, depicted as a irregular low density area on the T1 image, this can more clearly be seen as a high density area on T2 image (shown to the right in Fig. 4.36); this delineates the CTV. The CTV includes regions of subclinical tumor that is adjacent to the GTV; this is the new target of B/GdNCT. GTV can be treated by B/GdNCT. However, regions of the CTV are partially protected by the BBB that prevents the passage of most B/Gd compounds. To be effective, these compounds will

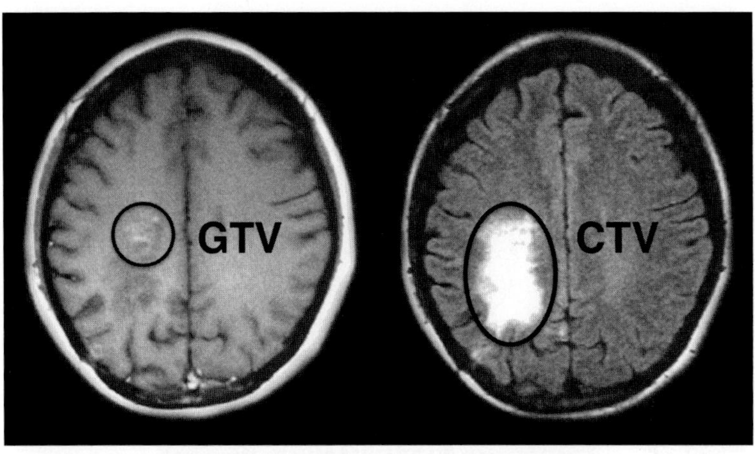

Figure 4.36 G-III glioma, anaplastis astrocytoma (*left*: T1 image enhanced with GdDTPA, GTV: gross tumor volume, *right*: T2 image, CTV: clinical tumor volume).

have to traverse the BBB, and the blood tumor barrier (BTB) making treatment by NCT difficult, since most of the B/Gd compounds cannot pass through normal blood-brain barrier (BBB). Thus ideally, the B/Gd compound should pass through the BBB and accumulate preferentially in tumor cells with long retention times and have rapid clearance from the normal parenchyma. To date, such compounds have not been identified.

Modification of the BBB in the peri-lesional rim or the peri-tumoral edema should be investigated for selective delivery of B/Gd agents. Most of the malignant brain tumors begin with malignant changes or dedifferentiation of the normal glial cells; they grow rapidly by accompanying angiogenesis to feed themselves. However, fortunately, the barrier between tumor parenchyma and neo-tumor vessels (BTB) is much more permeable than the BBB; for example, GdDTPA can pass through BTB, but not the BBB.

As shown in Fig. 4.37, the peri-lesional rim (the peri-focal edematous lesion) is the new target for NCT. Since the GTV can presently be cured via BNCT, new compounds must be developed to target this peri-lesional rim. Although tumor cells might be invading this region, along the neural matrix, the BBB is almost intact, while BTB is premature. This is a kind of no man's land for NCT. Fortunately, high-LET-ionizing γ radiation

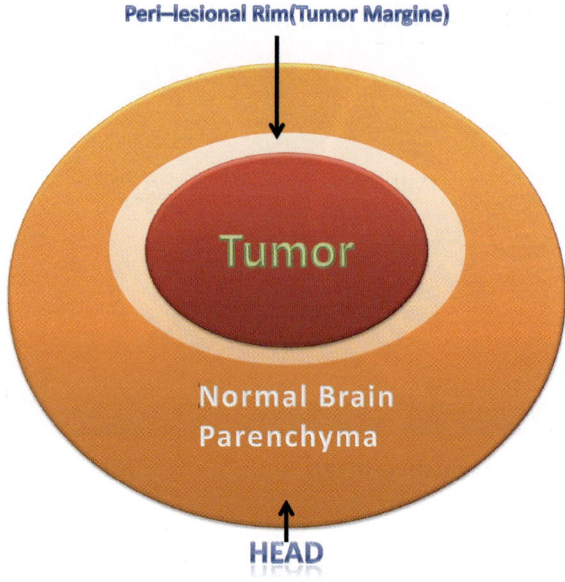

Figure 4.37   Schematic view of peri-focal edematous lesion as a peri-lesional rim.

component of GdNCT with long trajectories can transfer their energy onto the peri-focal edematous lesion beyond the GTV. Generally, it is very difficult to deliver any agent selectively to invading tumor cells in the peri-focal edema. There are strategies for drug delivery involving the destruction or modification of the BBB. However, the investigation of BBB-modification requires a great deal of circumspection. The BBB is maintained by very complicated mechanisms such as transporter systems. Only a few small molecules of high lipid solubility and low molecular mass can passively diffuse through the BBB.[90a] In order to deliver boron compounds, BBB disruption is one strategy, but its clinical application is highly controversial for safety reasons.[244]

Since gadolinium compounds have been used as enhancing agents for MRI, they are commercially available and have been thoroughly tested. Table 4.1 lists some the MRI agents, and Fig. 4.38 gives the structures of some of the Gd^{3+} complexing agents. While some of these, such as GdDTPA and GdDOTA, can pass through the BTB, they are not retained in tumor and are rather rapidly washed out via the urinary system.

Table 4.1 Commercially available Gd compounds.

Gd-Ligand	Chemical Name (Registered)	Chemical Formula	MW	Ionicity	Chem Struct	Since
Gd-DTPA	Meglumine Gadopentetate (Magnevist)	$C_{14}H_{20}GdN_3O_{10}C_7H_{17}NO_5$	742.79	Ionic	Linear	1988
Gd-HP-DO3A	Gadoteridol (ProHance)	$C_{17}H_{29}GdN_4O_7$	558.69	Nonionic	Macrocyclic	1994
Gd-DTPA-BMA	Gadodiamide (Ominiscan)	$C_{16}H_{28}GdN_5O_9$	645.72	Nonionic	Linear	1996
Gd-DOTA	Meglumine Gadotetate (Magnescope)	$C_{16}H_{25}GdN_4O_8C_7H_{17}NO_5$	753.86	Ionic	Macrocyclic	2001

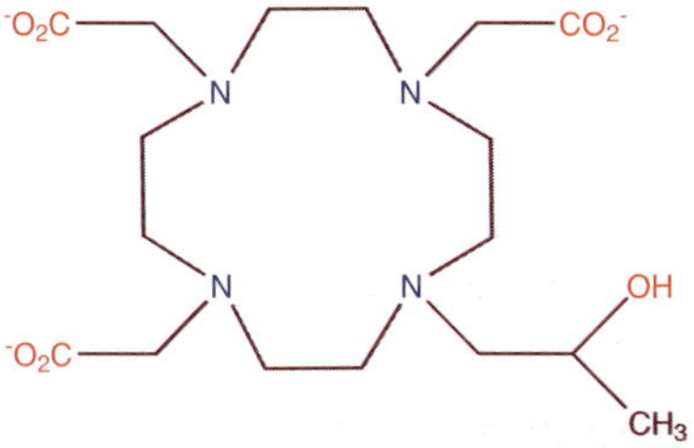

Gd-HP-DO3A

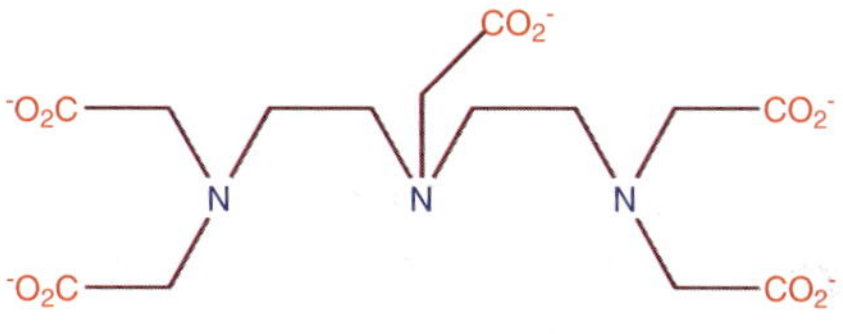

Gd-DTPA

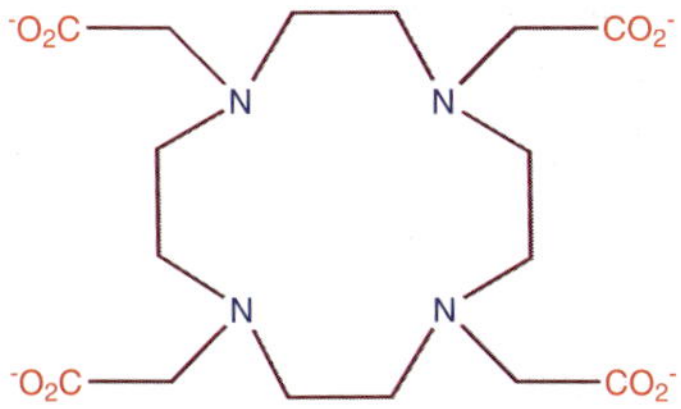

Gd-DOTA

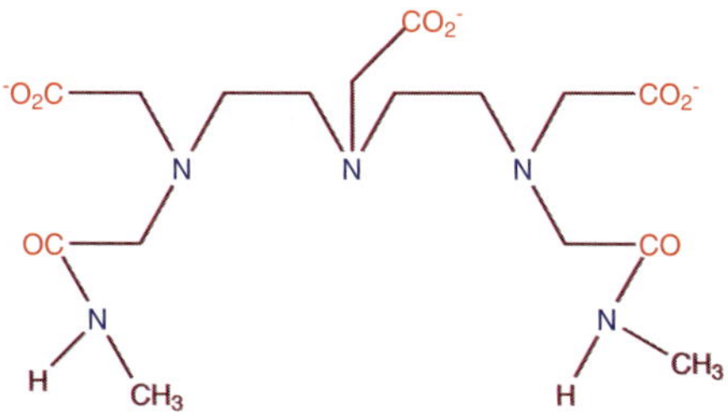

Gd-DTPA-BMA

Figure 4.38 Gd chelating agent to conceal its toxicity.

Although GdDTPA is taken up into tumor cells *in vitro*, it is not accumulated into human tumor cells *in vivo*. This is one of the pitfalls of *in vitro* GdNCT studies. Thus, GdDTPA is not utilized for GdNCT. However, good results have recently been obtained from GdNCT on tumor models via continuous intravenous injection of GdDTPA during GdNCT. This study revealed that GdNCT might be satisfactory to control tumor growth, even if the Gd atom does not bind directly to DNA. Large quantities of Gd can be readily introduced into tumors by direct intratumoral injections. Khokhlov *et al.* reported that GdNCT via intratumoral injection of GdDTPA showed significant antitumor effects from internal conversion electrons and γ-rays, even if Auger electron's effect was excluded.[245]

For clinical purposes, however, it will be essential to develop Gd agents that bind to DNA for the following reasons; (1) to minimize the dose of Gd agents, (2) to minimize shielding effects, and (3) to utilize the high-LET Auger electrons. However, it could take a long time to develop DNA-seeking Gd agents.

There have been attempts to attach Gd agents to nucleus-penetrating entities. One example is the bis(Pt(II)terpyridine)Gd(DTPA) complex, whose synthesis is shown in Scheme 4.5.[246] The cellular uptake of the Pt-Gd complex was studied *in vitro* using A549 human lung carcinoma cells and human aortic endothelial cells. It was found that the Pt-Gd DNA metallointercalator stayed intact and had a propensity to bind to DNA and selectively accumulate within the cancer cells.[246]

It should be pointed out that there have been a number of cases involving some medical side effects thought to be associated with some gadolinium contrast agents. In 2007, the US Food and Drug Administration (FDA) asked manufacturers to include a new warning on the product labeling of all gadolinium-based contrast agents. The warning states that patients with severe kidney insufficiency who receive gadolinium-based agents are at risk for developing a debilitating, and a potentially fatal disease known as nephrogenic systemic fibrosis (NSF). In addition, patients just before or just after liver transplantation, or those with chronic liver disease, are also at risk for developing NSF if they are experiencing kidney insufficiency of any severity.

Recently, there are a number of reports of Gd-containing nanoparticles.[247] Fluorescent nanoparticles containing a gadolinium oxide core

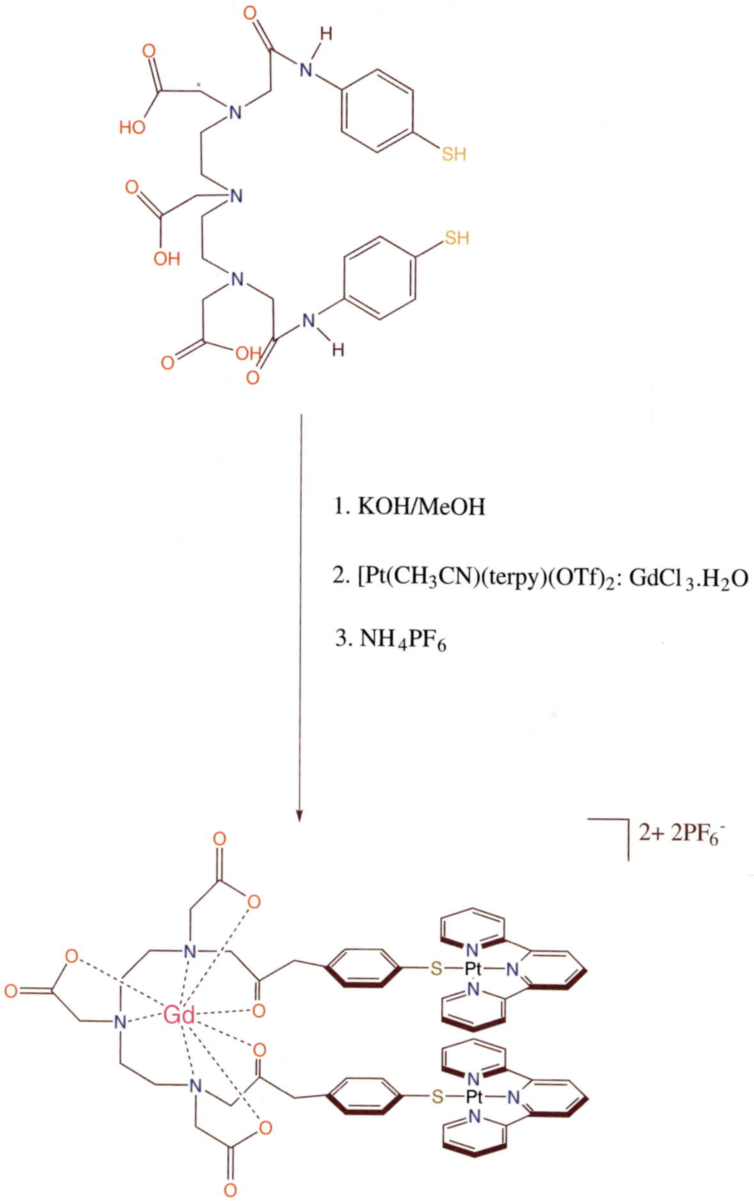

Scheme 4.5 Synthesis of bis(Pt(II)terpyridine)(GdDTPA) compound.

are very attractive because they are able to combine both imaging (fluorescence imaging, MRP) and therapy (X-ray therapy and neutron capture therapy) techniques. The exploitation of these multifunctional particles for *in vivo* applications will require accurate control of their biodistribution.

Novel submicrosized graphitic carbon shells embedding nanometric Gd(III) oxidic phases have been reported to be potential candidates for dual diagnostic (MRI) and therapeutic (neutron capture therapy) applications.[248] There have also been reports of small single-wall carbon nanotubes containing Gd^{3+} ions incapsulated in the interior of the nanotubes. These composites were found to be linear supermagnetic molecular magnets with MRI efficiencies 40 to 90 times larger than any known Gd^{3+} contrast agents; no biological studies were reported.[249]

Gadolinium-conjugated TiO_2-DNA oligonucleotide nanoconjugates with increased retention time, Gd accumulation, and intracellular delivery may find its use in Gd neutron capture cancer therapy.[250]

Instead of GdDTPA, gadobenate dimeglumine (Gd-BOPTA; see Fig. 4.39) showed significantly prolonged delay of tumor growth as compared to gadopentate dimeglumine (Gd-DTPA) via intratumoral injection into subcutaneously growing 9L gliosarcoma. The authors emphasized that Gd-BOPTA warrant further investigation of subcellular Gd distribution.[251]

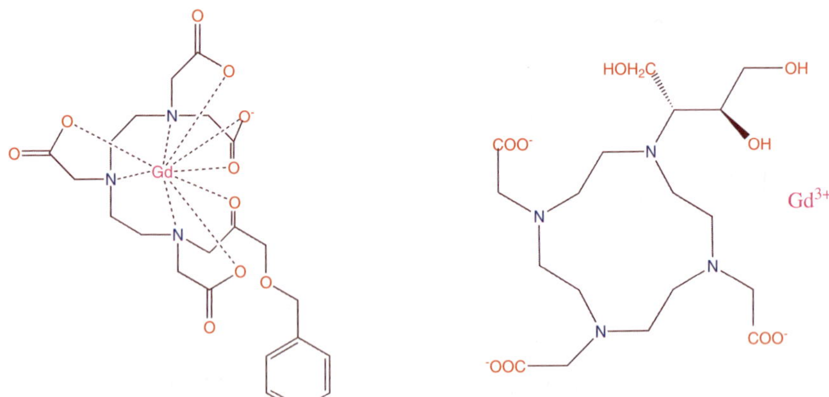

Figure 4.39   Formula of the contrast agents Gd-BOPTA and gadobutrol.

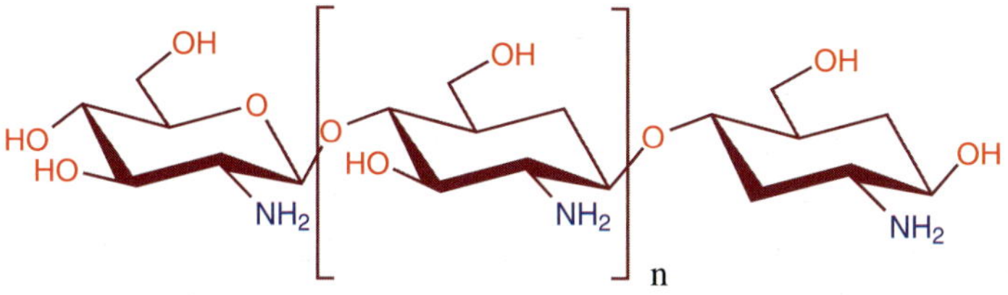

Figure 4.40 Structure of chitosan.

A neutral macrocyclic gadolinium complex, Gadobutrol (see Fig. 4.39) showed a therapeutic benefit for GdNCT via intratumoral injection and is a promising candidate for GdNCT.[252] In addition, chitosan (see Fig. 4.40) was reported to be a highly promissory ligand of GdDTPA for GdNCT.[198,254] Further preclinical studies should be done.

Lipid-based nanoparticles containing complexes of gadolinium have been studied as contrast enhanced MRI agents. Le *et al.* reported the encapsulation of a Gd compound into a liposome formulation followed by the dispersion of the liposome into a thermo-sensitive polymeric gel. In murine tumor models, they showed that this liposome-in-thermo-sensitive gel system significantly extended the retention of the Gd compound in tumors. Loposomal encapsulation will be a promising method for the delivery of Gd into tumor.[255] Le and Cui also reported that the Gd content in a tumor can reach an average of 159 μg/g of wet tumor tissue by encapsulating GdDTPA into a liposome. Using enriched forms of ^{157}Gd, this technique is already practical for GdNCT.[256] In related work, a gadolinium-containing lipid nanoemulsion, was reported to give a Gd concentration of 189 μg/g wet tumor which is adequate for GdNCT.[257] Furthermore, using gadolinium-incorporated lipid-nanoemulsions (Gd-nanoLE), Gd concentrations in a tumor reached 101 ppm Gd via intravenous injection, which was double that obtained via intraperitoneal injection. It has been estimated that between 50 and 200 μg of Gd/g of wet tumor tissue is required for effective GdNCT.[258]

Nanometer-scale chylomicron[259] emulsion and emulsifying wax and polyoxyl 2 stearyl ether nanoparticles containing high concentrations of gadolinium hexanedione (GdH)[260] are also reported to be effective for Gd delivery for GdNCT. Other types of ligands, such as folate,[261] chitosan,[262] thiamine,[263] Gd-DTPA-monoclonal antibody and Gd-porphyrins,[264] and avidin,[265] have been investigated for Gd carriers.

4.5. Combined Gadolinium and Boron Neutron Capture Therapy (GdBNCT) Agents

Gadolinium–boron complexes are highly unique and promising agents for future NCT. Conjugates of BSH and/or BPA combined with Gd-containing contrast agents can be used to quantify the amounts of boron compounds in tissues (see Fig. 4.41).

Complexes containing both Gd and B are of great interest both for therapy and for evaluating boron distribution using MRI during NCT. If MRI shows high Gd accumulation in a tumor, such compounds may also

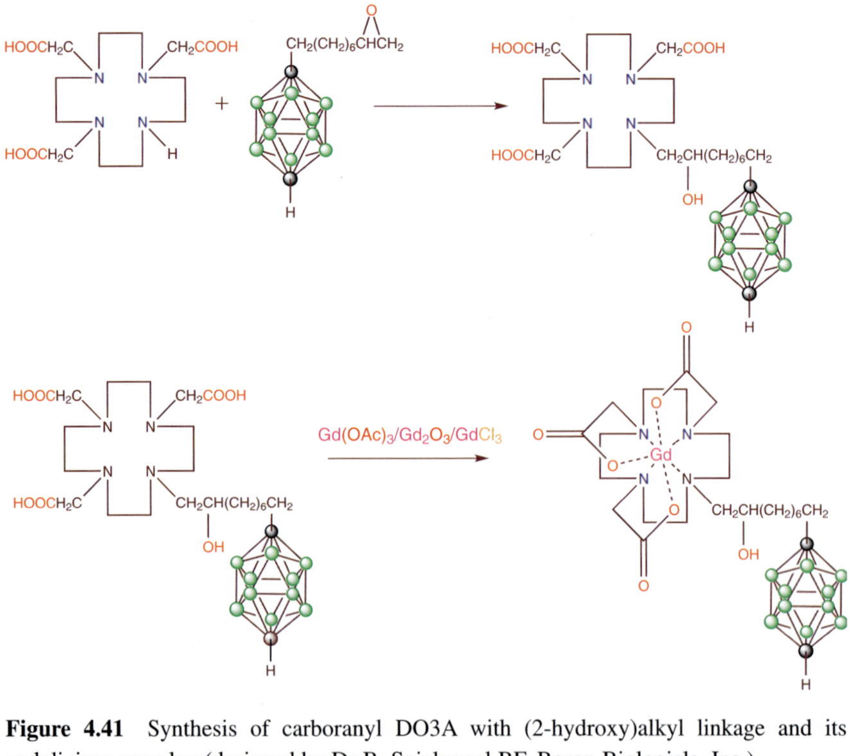

Figure 4.41 Synthesis of carboranyl DO3A with (2-hydroxy)alkyl linkage and its gadolinium complex (designed by Dr B. Spielvogel BF, Boron Biologicla, Inc.).

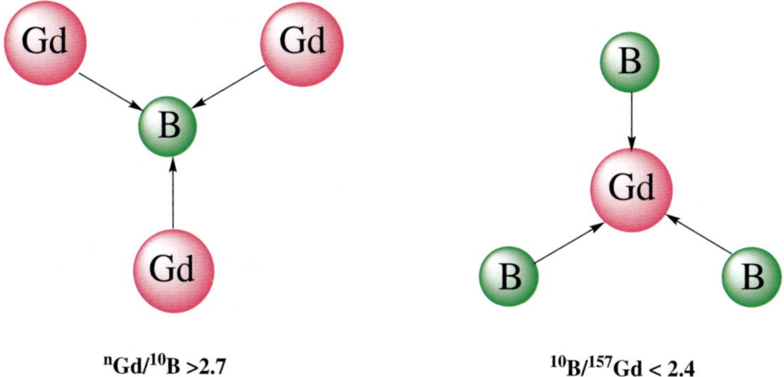

Figure 4.42   Therapeutic requirements of B–Gd compounds for effective GdNCT/BNCT.

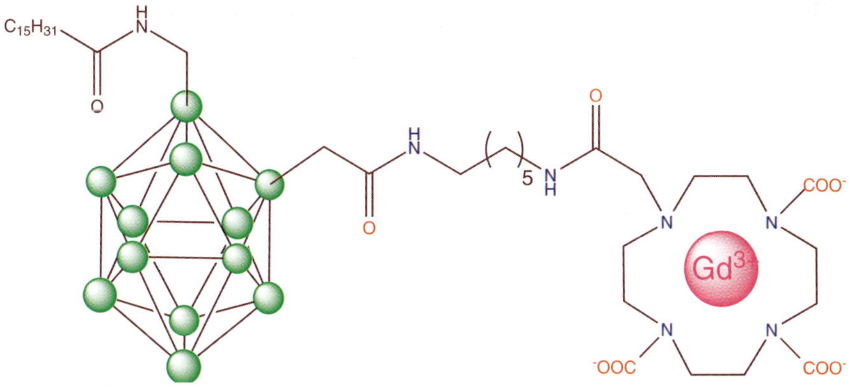

Figure 4.43   Structure of Gd(III)-*C*-(DOTAMA-C$_6$-amidomethyl)-*C*'-palmitamidomethyl-*o*-carborane.

help with diagnosis and decision making for Gd/BNCT. Preliminary studies showed that, if both ^nGd and ^{10}B atoms could be used for therapeutic effect, the optimal ratio of ^nGd to ^{10}B in tumors should be about 3. If the ratio $^nGd/^{10}B$ is smaller than 3 or the ratio of $^{10}B/^{157}Gd$ is larger than 3, as a rule, the Gd will not be effective in therapy, but will still be useful as a contrast agent in MRI (see Fig. 4.42).

If therapeutic concentrations of ^{10}B and ^{157}Gd in tumors are 30 ppm and 200 ppm, respectively, then the Gd/B atom ratios could be calculated as > 2.4 for ^nGd and > 0.43 for ^{157}Gd. The lower ratio for ^nGd would make such compounds less effective for therapy, but still useful as a diagnostic contrast medium. For example, the Gd(III)-*C*-(DOTAMA-C_6-amidomethyl)-*C'*-palmitamidomethyl-*o*-carborane, shown in Fig. 4.43, has been synthesized as a combined BNCT/MRI contrast agent.[266] No toxicity or biological distribution studies have been reported.

Combinations of both BNCT drugs and GdNCT drugs have been studied *in vitro* in Chinese Hamster fibroblast V79 cells.[267] The neutron targeting compounds, ^{10}BSH at 0, 5, 10, and 15 ppm and ^{157}Gd (GdBOPTA) at 0, 800, 1600, 2400, 3200, and 4800 ppm were combined, followed by nuclear irradiation. The combinations of B and Gd drugs showed an additive effect at Gd concentrations below 1600 ppm. The additive effect decreased as a function of the gadolinium concentration at 2400 ppm and showed no additive effect at more than 3200 ppm of Gd. The results showed an enhanced Gd effect at low Gd concentrations up to a certain ratio, after that the large capture cross section of Gd effectively shields the boron compound from neutron capture. Such behavior would most probably be found in a single Gd/B drug.

Thus, a number of promising boron and gadolinium delivery agents have been synthesized and preliminary *in vitro* and *in vivo* results have been obtained. However, these stop mainly with element distribution studies, in mice and small test animals. Little is known about the effectiveness of these compounds in clinical NCT studies. Despite the elegant procedures and rationalizations for their syntheses, there are still only two BCNT drugs currently in clinical trials, BSH and BPA. Although these have been found to be safe, they are not very effective. More compounds must be taken to clinical trials. However, in view of what this entails, in terms of facilities, it will be a very slow process. Even fewer results are available for GdNCT. Most compounds tested are MRI contrast/enhancement agents and these agents are short-lived in the body; without continuous administration their use in GdNCT will be limited.

Chapter 

Neutron Sources for NCT

5.1. Introduction

Any neutron source for BNCT must furnish a beam of neutrons of the proper energy and of sufficient intensity so that treatment time is less than 1 hr, preferably not more than 30 min. Generally, neutrons can be classified according to their energies as thermal ($0 < E < 0.4$ eV), epithermal $0.4 < E < 10$ keV), and fast neutrons ($E > 10$ keV). It is the low energy, thermal neutrons that are therapeutically important. However, thermal neutrons do not penetrate tissue to any great extent. At a depth of about 2.5 cm below the skin surface, such a beam flux is decreased by ~50%. Thus, thermal neutrons would only be effective for treating surface tumors in a BNCT treatment. Poor neutron penetration, along with the nonspecificity of the boron carriers used, was the major reason for the failures of the earlier (1950–1960) clinical trials in the USA.[71] Generally, what is done in the treatment of deep-seated tumors is to use higher energy, epithermal, beams which lose much of their energy (become thermalized) in the vicinity of the tumor.

5.2. Fission-Based Reactors

Although one of the first BNCT experiments used a cyclotron as the neutron-generating source,[269] most of the clinical trails now in effect use

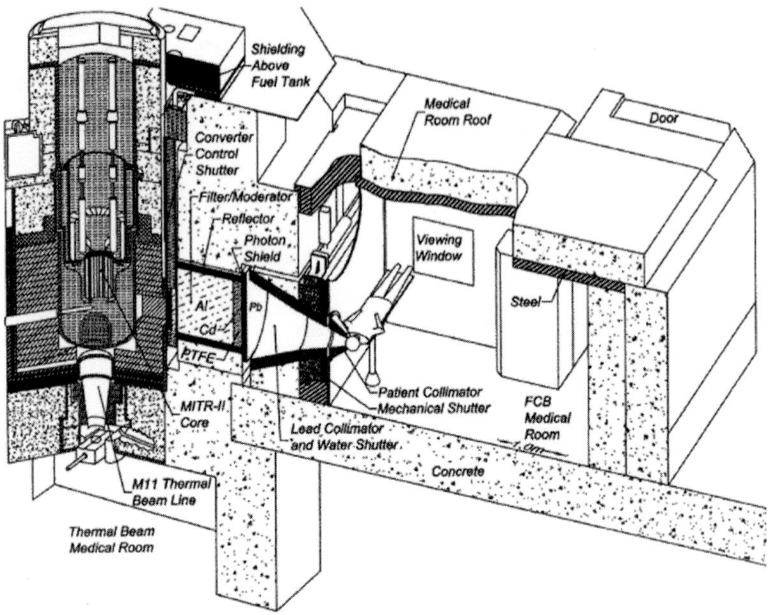

Figure 5.1   Cutaway drawing of the MITR-FCB epithermal reactor.[268]

nuclear reactors as their neutron source. Harding and Riley have recently reviewed fission reactors as neutron generators and the reader is referred to that review.[268] Figure 5.1 shows a diagram of the MITRR with the epithermal fission converter beam (FCB), and it is described elsewhere in more detail.[269] An epithermal beam from the reactor with a 16 cm diameter circular collimator aperture, has an intensity of 5×10^9 n cm^{-2}s^{-1} with a contamination of 1.2×10^{-13} Gy cm^2 and 3.2×10^{-13} Gy cm^2 for fast neutrons and gamma rays, respectively.[268] The Gy (**gray**) is the SI unit of absorbed radiation dose of ionizing radiation and is defined as the absorption of 1J of ionizing radiation by 1 kg of matter. This same reference lists some characteristics of an ideal beam and is reproduced in Table 5.1.

Since one of the main potential uses of BNCT is the treatment of brain tumors, such as glioblastoma multiforme, in which infiltration to normal brain tissues is a key factor, large volumes of the brain must be treated.

Table 5.1 Desirable properties of neutron beams for BNCT.

	Epithermal Neutron Beams	Thermal Neutron Beams
Energy	0.4 eV $< $ E < 10 keV	$0 < $ E < 0.4 eV
Intensity	$> 2 \times 10^9$ ncm^{-2}s^{-1}	$> 0.5 \times 10^9$ ncm^{-2}s^{-1}
Contamination	$< 2.8 \times 10^{-13}$ RBE Gy cm^2 for gamma and fast neutron	$< 2.8 \times 10^{-13}$ RBE Gy cm^2 for gamma and fast neutron
Collimation(J/φ)	> 0.75	—
Advantage depth	> 9 cm	> 4 cm
Beam size	0 to > 16 cm diameter	0 to > 16 cm diameter

The role of a neutron source is to create an efficient thermal neutron field in the boron-labeled tumor cells within the targeted volume; this is accomplished by using a beam with a range of neutron energies. There are several figures of merit that have been used to assess the quality of a neutron beam. Figure 5.2 shows a dose-depth curve for the MIT-II reactor, in a head phantom for a 12 cm field. The curve assumes BPA as the boron source with a tumor/brain boron ratio of 3.5. The *advantage depth* (AD) for this beam is 8.9 cm. This is defined as the depth in the tissue at which the dose to the tumor (therapeutic dose) is the same as the dose to the brain (background dose); it is a measure of the maximum useful depth for any therapeutic benefit. The AD depends on how free the beam is of contaminating photons, thermal and fast neutrons, and the tumor/brain boron ratio. In general, the AD increases with neutron beam energy up to about 10 keV after which there is a sharp drop-off; the decrease is due to the energies produced by recoil protons.[272] An ideal epithermal beam is one that is free of contamination and whose neutron intensity varies as 1/E in the epithermal region; there are neutron beams that closely approximate these optimum beam characteristics.[269] The *advantage ratio* (AR) is defined as the integrated dose delivered to the tumor, up to the advantage depth, divided by the integrated dose deliver to the tissue, along the central treatment axis. For the example shown in Fig. 5.2, the advantage ratio is equal to 5.0. The *advantage depth dose rate* (ADDR) is the maximum Relative biological effectiveness (RBE) dose rate delivered to total tissue, or the RBE dose rate to the tumor at the AD.

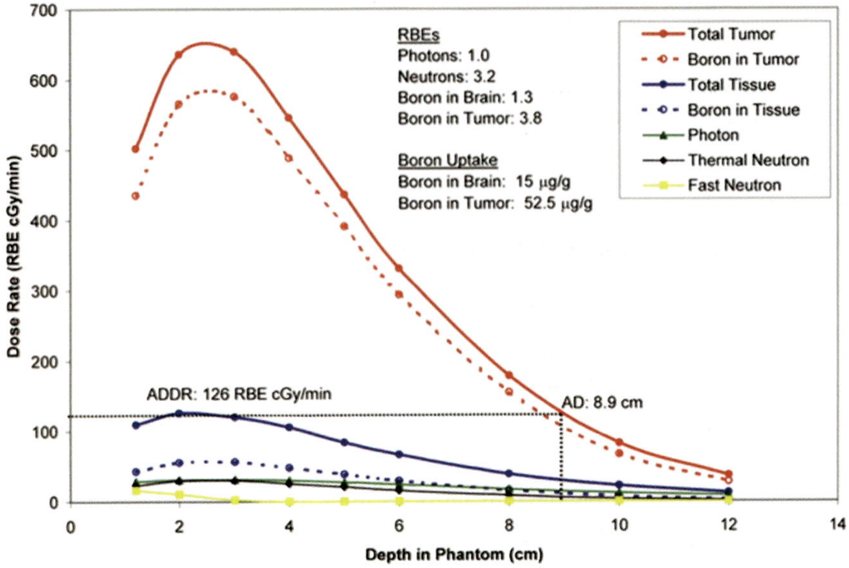

Figure 5.2   Typical depth-dose curve for the MITR-FCB reactor using BPA.[271]

The depth-dose curves for different reactor beams can be calculated using Monte-Carlo techniques, such as MCNP.[273] Treatment plans, such a *SERA: Simulation Environment For Radiotherapy Applications*, have been developed for BNCT,[274] other plans include NCTPlan,[275] and THORPlan.[276]

Body phantoms have been developed for dosimetric studies. Since BNCT is mostly involved with brain tumors, the phantoms normally model the brain (head phantom). In BNCT, the head phantom should be made of materials that have similar neutron and photon scattering and absorption properties as found in the human head. Water-filled phantoms with polymethylmethacrylate (PMMA) sides, PMMA, polyethylene (PE), gels and tissue equivalent (TE) liquids have been used.[277,278] The TE liquids are aqueous solutions whose elemental compositions mirror those found in the brain.[279] A comparison of the different tissue equivalents shows that water is an acceptable brain tissue substitute for periodic epithermal beam calibration, for routine quality control, and for intercomparison of different beams in NCT.[279] When high accuracy is sought, one of the TE liquids is preferable.[280]

It is estimated that for high throughput clinical BNCT an epithermal beam intensity of $\geq 2 \times 10^9$ n cm^{-2}s^{-1} with an energy spectrum of 0.4 eV $<$ $E > 10$ Kev and a contamination of $< 2.8 \times 10^{-2}$ Gy cm^2 per neutron per cm^2 is required.[281] At the present time, nuclear reactors are the only sources for such a neutron beam.

5.3. Accelerator-Based Neutron Sources

The currently required use of nuclear reactors as neutron sources continues to restrict the testing of new compounds and delivery systems for BNCT. While there are some reactors that are being developed specifically for medical use, it is unlikely that they will be located in a hospital setting. If BNCT is going to be readily available, other neutron sources must be developed. One obvious alternative is the use of a particle accelerator to generate the necessary neutron flux. The advantages of an accelerator-based neutron source (ABNS) are that they would deliver a higher quality neutron beam than an idealized standard reactor neutron beam (ISRNB),[282] and they could be installed in a hospital environment at a reasonable cost. Table 5.2 lists some particle reactions that could be used in a possible ABNS.[282] It should be noted that all of the neutron beams in Table 5.1 are too high in energy and they would require moderation to the epithermal range (0.4 eV–10 keV). The most considered is the ^{7}Li$(p,n)^7$Be reaction. This produces a lower energy beam that would require the least moderation. A smaller moderator length would mean a reduced neutron loss and, hence, delivery of a higher neutron beam flux at the target. The threshold proton energy for the ^{7}Li$(p,n)^7$Be reaction is ~1.89 MeV. Calculations by Zimin and Allen[283] showed that neutron beams generated in the near threshold proton energy range would require moderation and that there is little gain in useful neutron flux between 1.89 MeV and 1.95 MeV proton beams. However, there was a sizeable increase in neutron flux in going to 2.5–3.0 MeV protons. Most simulations and calculations are done on ABNS's based on the ^{7}Li$(p,n)^7$Be reaction using 2.5 MeV proton beams. In general, the higher the proton current, the greater the neutron flux and the shorter the treatment time. Accelerators can be classified as being linear or circulating, and both types have been explored.

Table 5.2   Charged particle reactions as potential neutron sources.[282]

Reaction	Bombarding Energy (MeV)	n Production Rate (n/min.-mA)	Average 0° n Energy (MeV)	Maximum n Energy (MeV)	Target MP (°C)
^{7}Li$(p,n)^7$Be	2.5	5.3×10^{13}	0.55	0.79	181
^{9}Be$(p,n)^9$B	4	6.0×10^{13}	1.06	2.12	1287
^{9}Be$(d,n)^{10}$B	1.5	1.3×10^{13}	2.01	5.81	1287
^{13}C$(d,n)^{14}$N	1.5	1.1×10^{13}	1.08	6.77	3550

A cyclotron is an example of a circulating accelerator. They have been used in a hospital setting for the production of positron-emitters and for fast neutron therapy.[282] However, the current of 10 mA of 2–2.5 MeV protons is well outside the capability of most standard hospital cyclotrons. However, Tanaka *et al.* have reported simulations based on epithermal neutrons generated by a 1 mA 30 MeV cyclotron generated proton beam and a Be target (see Table 5.1) that is planned for Kyoto University Research Reactor Institute (KURRI).[284] Based on a 13 ppm concentration of ^{10}B in normal brain tissue an irradiation time of less than 30 min would result in a dose of less than the suggested maximum of 12.5 Gy-eq to normal tissue. In addition to brain tumor research, the KURRI accelerator has been studied for multiple liver tumors and malignant pleural mesothelioma.[285]

Most of the accelerators described in the literature for BNCT are linear accelerators (Linac). There are several types, electrostatic accelerators in which a beam of charged particles (protons) is accelerated by a high potential to impinge on a selected target, such as those listed in Table 5.2. In the so-called tandem electrostatic accelerator, a stream of hydride (or deuteride) ions are accelerated by a high positive potential and then stripped of their electrons to yield protons which are repelled by the same positive potential. This has the advantage that a single potential can accelerate a hydrogen twice, first an attractive force acting on the H$^-$ ion, then a repulsive force acting on the proton. They also offer the advantage that both the source and target are at ground potential.[282,286] Tandem-Electrostatic-Quadrupole accelerators that can deliver 30 mA of 2.5 MeV protons used with Li targets have also been described.[287] One of the major technical problems associated with all accelerator-based neutron sources is that of heat removal. There are two methods of heat removal generally used, submerged jet impingement of a cooling fluid and the use of microchannel vents to enhance radiant cooling. Light and heavy water as well as liquid gallium have been used as the cooling fluid.[282] There is always a trade-off between neutron beam flux and target choice. According to Table 5.2, ^{7}Li would be the preferred target, neutrons produced would require less moderation and hence be subject to less neutron beam attenuation. However, from a physical stand point, Li is the least attractive, it has a low melting point and lower thermal conductivity, in addition, the ^{7}Be product is radioactive, which introduces its own complications. Replacing the Li metal with a refractory lithium compound, such as Li_2O (MP 1500°C) would cut down the neutron flux by ~50%.[287] A thin layer of Lithium is deposited on a metal surface, which is cooled by circulation of water through microchannels on the reverse side of the surface.[288]or through cooling loops in the target assembly.[289] Conical-shaped heat exchangers with water channels on the outside have also been studied.[290] These efficient heat removal systems can maintain Li below its melting point with proton beams sufficient to product the required neutron fluxes ($\sim10^9$ n cm^{-2} s^{-1}) for BNCT.[291]

In view of the problems associated with using lithium, other targets have been considered. The ^{9}Be(p,n)^{9}B has a higher neutron yield, the target has an extremely high melting point (1287°C), and nonradioactive products are formed. Suzuki *et al.*, compared irradiation times and dose distributions for the newly constructed accelerator-based neutron generator (AB-BNCT) at KURRI with the KUR reactor neutron generator (RB-BNCT). Treatment plans were constructed for multiple liver tumors and malignant pleural mesothelioma. The AB-BNCT system consisted of a cyclotron accelerator that delivered a proton beam of ~2mA at 30 MeV, with a Be target. Dose distribution analysis revealed that AB-BNCT was superior to the RB-BNCT for deep-seated tumors.[285] Deuterium-induced nuclear reactions, including ^{9}Be(d,n)^{10}B, ^{12}C(d,n)^{13}N, and ^{13}C(d,n)^{14}N, have been considered (see Table 5.2).[292] On the basis of a high yield of low energy neutrons, the ^{13}C(d,n)^{14}N was argued to be the most promising.[292]

5.4. Other Neutron Sources

5.4.1. *Fusion-based neutron generators*

This neutron generator is an accelerator-based system, but based on a $^3H(d,n)^4He$ (D-T) or a $^2H(d,n)^3He$ (D-D) fusion reaction. A series of compact neutron generators are being developed at Lawrence Berkeley National Laboratories (LBNL). They utilize a 13.4 MHz RF induction discharge to produce a deuterium or deuterium–tritium beam that is accelerated (100–300 kV) toward a titanium target to produce the desired fusion reaction.[293,294] One advantage of these generators is their small size, a coaxial neutron generator that can deliver 10^{10} ns^{-1} based on a D-D reaction is 30 cm in diameter and 40 cm in height.[294] Deuterium–tritium reactors can product a higher neutron yield, but suffer from the use of radioactive tritium. It is also possible to use fission multipliers to increase the nuclear flux.[295] These might well be the neutron sources that will eventually be based in hospitals for BNCT.

5.4.2. *Spallation-based systems*

A spallation process is the generation of neutrons by the high energy bombardment of protons (30–600 MeV) on a metal target, such as tantalum, copper, or mercury. Figure 5.3 shows a schematic of a spallation neutron generator using a tantalum as a target.[296] The neutrons produced by this process are very high in energy and they must be heavily moderated. The schematic in Figure 5.3 shows the moderator (Fe) thickness, the beam shaping filter (AlF_3/Al/LiF), and the Pb reflector for a 50 MeV proton beam and a Ta target. Such beams can be generated by commercially available cyclotrons. However, there are no known plans for the construction of a spallation-based BNCT facility.

5.4.3. *Isotope decay processes*

Isotopic neutron sources offer a completely different approach to BNCT. Californium-252 ($t_{1/2}$= 2.645 yr) is a spontaneous fission neutron source that can furnish an intense neutron beam (^{252}Cf decay rate = 2.314 $\times$ 10^9 $ns^{-1}mg^{-1}$). Most of the isotope is produced in the high flux reactor at

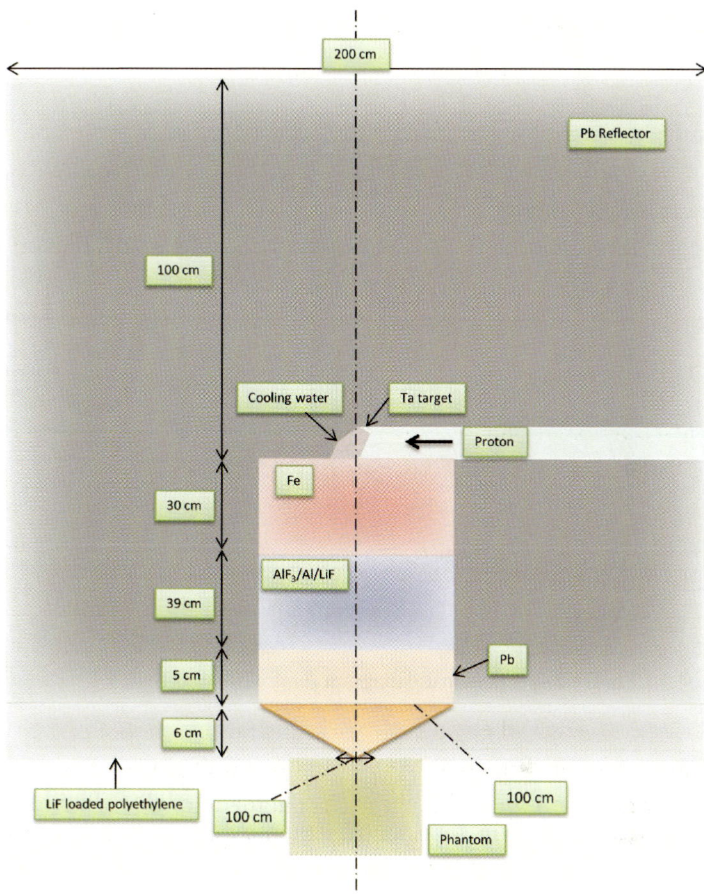

Figure 5.3   Schematic diagram of a spallation-based neutron source for BNCT.[297]

Oak Ridge National Laboratories. The production process is quite lengthy, it requires an eight months reactor time followed by a four-month period for cooling and radiochemical processing. The yearly yield is about 100 mg.[297] It has been used extensively in both industry and medicine as a portable neutron source. Figure 5.4 shows a proposed schematic diagram of a BNCT-based apparatus based on ^{252}Cf decay.[298]

Unfortunately, the amount of ^{252}Cf required to give a neutron flux sufficient for BNCT (~100 mg) is about the total amount currently produced

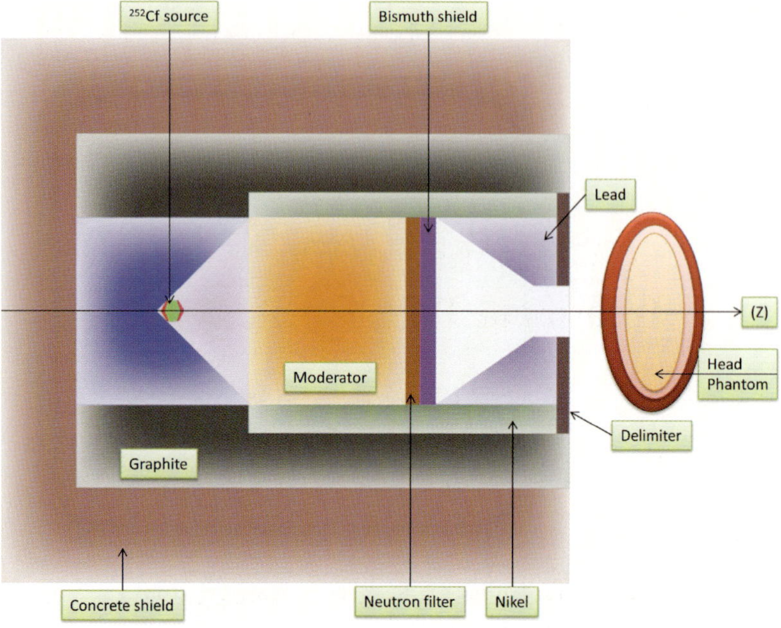

Figure 5.4.   Schematic diagram for ^{252}Cf-based BNCT.[298]

in the USA in a year.[299,297] On the other hand, ^{252}Cf has been extensively used as a point source in fast neutron therapy for cervical cancer. The Cf^{252} can be implanted at the surface of the tumor, which cuts down on the dose to healthy tissue.[300] The combination of ^{252}Cf brachytherapy enhanced by boron neutron capture (BNCEFNT) could product an enhancement of therapeutic gain by as much as 30%.[301] In the case of fast neutron therapy, for brain cancer and melanoma, such an enhancement could increase cell-kill between 10 and 100 fold.[302]

Chapter 6

NCT Dosimetry and Treatment Planning

6.1. BNCT Dosimetry

Dosimetry is the measurement of ionizing radiation coming in contact with matter, specifically body tissue. In BNCT, the process is more complex than that found in photon and electron radiation.

There are four sources of radiation: (1) the dose due to ^{10}B fission from the ^{10}B$(n_{th},{}^4_2\text{He})^7$Li reaction; (2) neutron dose due to the recoiling protons from the epithermal and fast neutrons ^{1}H(n, n')H$^+$ reaction; (3) gamma rays from gamma ray contamination of the beam and the ^{1}H$(n, \gamma)^2$H reaction that emits a 2.2 MeV gamma ray; and (4) proton dose from the ^{14}N$(n, p)^{14}$C reaction which releases 0.626 MeV of energy which is deposited by the proton and recoiling C atom. In the ^{10}B reaction, the recoiling Li and alpha particle possess about 2.33 MeV of energy that is deposited within a distance of 5–9 µm, which kills the cell in which they are generated. About 94% of the time the recoiling Li ion is in an excited state and emits a 447 keV photon, which is usually ignored for therapy purposes, but can be used in ^{10}B analysis. The only therapeutic advantage is the ^{10}B fission dose, but only in as much as the ^{10}B concentration in the tumor exceeds that in normal tissue; the other radiation sources limit the extent to which BNCT is beneficial. The epithermal and fast neutron doses can be measured in air or in phantom. The former measurement is useful in characterizing the beam, while the latter is necessary in constructing treatment plans. As pointed out in the previous chapter, the phantom material is chosen to duplicate, as much as possible, the elemental composition of the tissue being irradiated.[251] Phantoms can be cylindrical or rectangular, made of solid PMMA or PMMA containers filled with water. An epithermal neutron beam will always be accompanied by gamma rays, thermal neutrons and fast neutrons, which must be measured. Both the $^{1}H(n, \gamma)^{2}H$ and the $^{14}N(n, p)^{14}C$ reactions are activated by thermal neutrons. These two reactions are quite important in limiting the radiation time in BNCT. Although ^{10}B has a much larger neutron cross section than either ^{1}H or ^{14}N ($\sigma_{th} = 3838$ b(^{10}B), 0.333 b(^{1}H), 1.83 b(^{14}N)), its concentration in a cell is, at best, orders of magnitude lower than either H or N in tissue.

Generally, gamma and neutron dose components are measured by a dual ionization chamber or dual dosimetry method. In this, readings are obtained from a tissue equivalent chamber that is equally sensitive to both photons and neutrons and from a graphite-walled chamber that measures only gamma rays. After suitable corrections, the difference in the two readings will give the neutron and photon doses in the beam. The thermal neutron flux is measured by a dual foil method in which readings are taken on a gold foil undergoing $^{197}Au(n_{th}, \gamma)^{198}Au$ reaction and on another gold foil covered by a layer of cadmium. Cadmium has a large capture cross section (~20,000 b) at thermal energies, but drops off rapidly at neutron energies above 0.1 eV, the so-called cadmium cut-off energy. The measurements of these doses in air are functions of the neutron beam and are used in monitoring the quality of beams. The in-phantom measurements are used for developing patient treatment protocols. Rogus, Harling and Yanch have described this method in detail for the MITR-II research reactor and have given estimated uncertainties in the calculated doses.[303] The total dose (D_{tissue}), sometimes called the equivalent dose (H_i), to healthy and tumor tissues at a particular point in the tissue is given by the relation:

$$D_{tissue} = D_H (RBE)_H + (D_\gamma + D_{b\gamma}) (RBE)_{photon}$$
$$+ D_{14_N} (RBE)_{14_N} + D_{10_B} (CBE)_{10_B} C_{10_B}. \tag{6.1}$$

Values for D_{14_N} and D_{10_B} are calculated from their kerma values.[304] The kerma (kinetic energy released in matter) is the expectation value of the energy transferred to charged particles at a point per unit mass. For neutrons, the D_{14_N} and D_{10_B} are given by the relations:

$$D_{14N} = \Phi_{th}\, F_n^{14_N}\, X_{14_N}$$
$$D_{10B} = \Phi_{th}\, F_n^{10_B},$$

(6.2)

where Φ_{th} is the neutron fluence rate in $n\ cm^{-2}\,s^{-1}$, $F_n^{14_N}$ is the kerma factor for ^{14}N in Gy-cm^2/n (or cGy-cm^2/n), $F_n^{10_B}$ is the kerma factor for ^{10}B per 1 ppm of ^{10}B in the tissue, and X_{14_N} is the weight fraction of nitrogen in tissue (brain = 0.022). D_H is the absorbed proton dose due to the $^1H(n,\,n')H^+$ reaction, also known as the fast neutron dose. D_γ is the induced absorbed gamma dose and $D_{b\gamma}$ is the gamma dose due to beam contamination. The RBE multipliers are the Relative Biological Effectiveness. The RBE of a particular test radiation can be defined as the dose of a reference radiation (usually X-ray) required to produce the same biological effect as the test radiation.[305] In the case of the dose due to ^{10}B, the Compound Biological Effectiveness (CBE) is used instead of RBE.[306,307] This is due to the unequal distribution of ^{10}B in a particular cell, which is both compound and tissue dependent. For example in rat skin studies, the CBE of BSH was 0.56 ± 0.06 for epidermal cell and BPA was 3.74 ± 0.7, while in vascular endothelial cells, the values were 0.86 ± 0.08 and 0.73 ± 0.42, respectively.[306] Using a rat spinal cord model, the CBE for BSH was 0.46 ± 0.5 compared to 1.33 ± 0.16 for BPA; the difference was rationalized by the fact that BSH cannot cross the BBB which protects the central nervous system (CNS) parenchyma, but BPA can cross the BBB and distribute in the CNS parenchyma.[307] It is recommended that the microdistribution of ^{10}B in tissue, using techniques such as ion microscopy be used in any therapy planning.[308] Equation (6.1) is location dependent; therefore, spatial dose distribution studied in phantoms are required. A number of gel dosimeters, such as Fricke-xylenol-orange gel dosimeters,[309] have been developed. This dosimeter is based on the Fricke solution dosimeter,[310] which has been known for a number of years. It is a solution of ferrous ammonium sulfate, sulfuric acid, and sodium chloride. In an acid solution radiation causes the decomposition of water according to the reaction:

$$H_2O \longrightarrow H + OH.$$

In a solution containing Fe^{2+}, this initiates the reaction sequence:

$$H + O_2 \rightarrow HO_2$$
$$Fe^{2+} + HO_2 \rightarrow Fe^{3(+)} + HO_2^-$$
$$HO_2^- + H^+ \rightarrow H_2O_2$$
$$Fe^{2+} + OH \rightarrow Fe^{3(+)} + OH^-$$
$$2Fe^{2+} + H_2O_2 \rightarrow 2Fe^{3+} + 2OH^-.$$

The Fe^{3+} generated can be measured either spectrophotometrically or by NMR from which the radiation dose can be calculated. The liquid Fricke dosimeter can be improved by dispersing the Fricke solution in a gel matrix, such as gelatin or agarose, and adding a metal complexing agent, such as xylenol orange (see Fig. 6.1). Such a dosimeter is capable of providing a spatial profile of radiation in a phantom. Even with a gel matrix, with a complexing agent, there is diffusion of iron ions over time which sets a practical limit on the time in which an irradiated Fricke gel can be imaged. On the other hand, polymer gel layer dosimeters (PGLD)[311] do not suffer from this limitation. These dosimeters are composed of thin layers of dispersed monomers that undergo polymerization under irradiation. In phantom dosimeters composed of activation foils, thermoluminescent detectors and ionization chambers have also been used to profile neutron beams.[312]

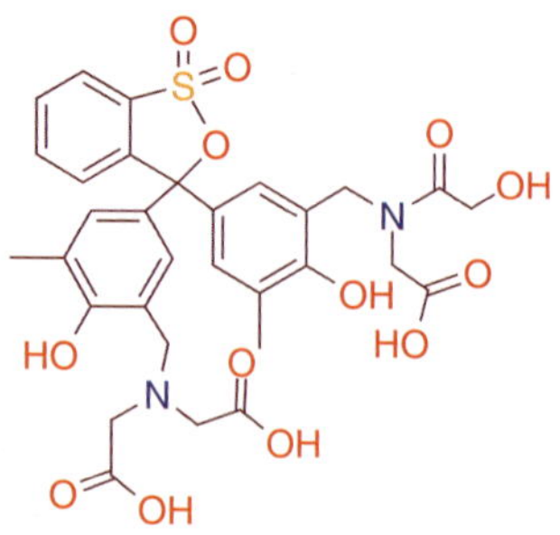

Figure 6.1   Structure of xylenol orange.

Beam characterization is the first step in constructing a treatment plan for specific patients. There has been an evolution of computer-based treatment plans. The geometries for neutron sources are complex. Fast neutrons emerging from the core must be moderated to the epithermal range, filtered and collimated. Different calculation methods are used for the different steps. Both deterministic methods, such as those found in TORT,[313] and stochastic methods such as contained in the various Monte Carlo codes, are used. Both have their advantages, but those based on Monte Carlo methods, such as found in *SERA*,[274] NCTPlan,[174(a)] THORPlan[176] and MacNCTPLAN,[174(b)] are far more numerous. There have been recent discussions regarding the advantages of combining NCT with other forms of therapy, such as external beam photo therapy.[315] This has lead to the development of treatment plans, such as MINERVA, for patients undergoing combinations of radiotherapy.[315]

Figure 6.2 shows 3-dimensional (3D) simulation images of the absorbed dose distribution as iso-dose curves, and Fig. 6.3 shows the 3D brain tumor depiction for decision making of positioning of patient's head. The absorbed dose is calculated as a summation of energy transfers from $^{10}\text{B}(n, \alpha)^7\text{Li}$,

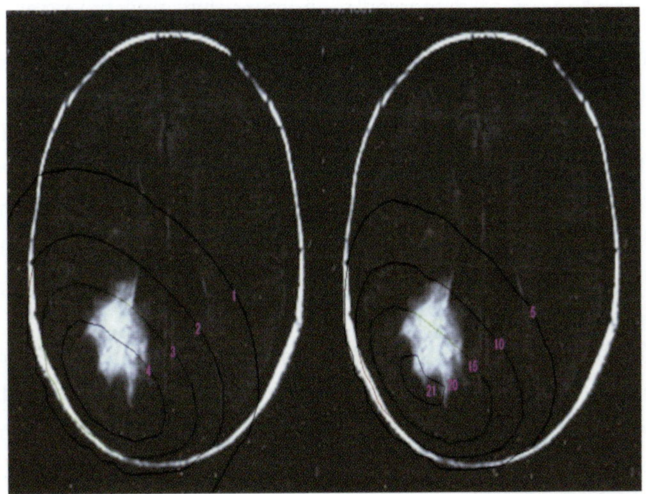

Figure 6.2   Dose equivalent curve to determine the irradiation period.

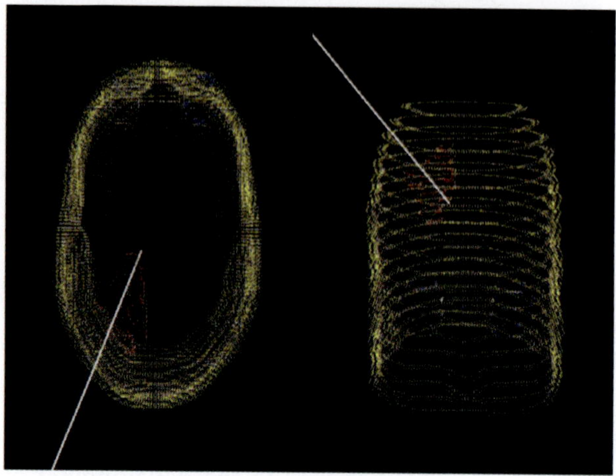

Figure 6.3   Three-dimensional simulation for irradiation planning.

^{14}N(n, p)^{14}C, and γ-rays, as shown in Eq. 6.1. Monte Carlo simulation methods are especially useful in studying thermal neutron random behavior in fluid and cellular structures. As shown in Fig. 6.4, the absorbed dose is strongly dependent on the boron microdistribution and the thermal neutron distribution.

Collimated thermal neutrons, incident onto the brain, targeting the brain tumor (Fig. 6.4 (b)). Thermal neutrons generate kinetic energy randomly, reacting with various atoms in the brain and all of the kinetic energy is absorbed in the tumor. The tumor-absorbed dose is estimated by the total kinetic energy of the ^{10}B(n, α)^{7}Li, ^{14}N(n, p)^{14}C, and contaminating γ-rays. The absorbed dose in the tumor is enhanced by particles and recoil Li ions emitted from ^{10}B atoms that selectively accumulate in the tumor (Fig. 6.4 (a)). Thus, the total selective destruction of the tumor is dependent on the boron microdistribution.

For the treatment planning, the boron concentration in the tumor, normal brain, and blood must be determined. Figure 6.5 shows the boron distribution after intravenous administration of BGPA (highly hydrophilic form of borono-glycyl-phenylalanine) to a brain tumor-bearing rat. The distribution is determined by a autoradiography, in which boron distribution

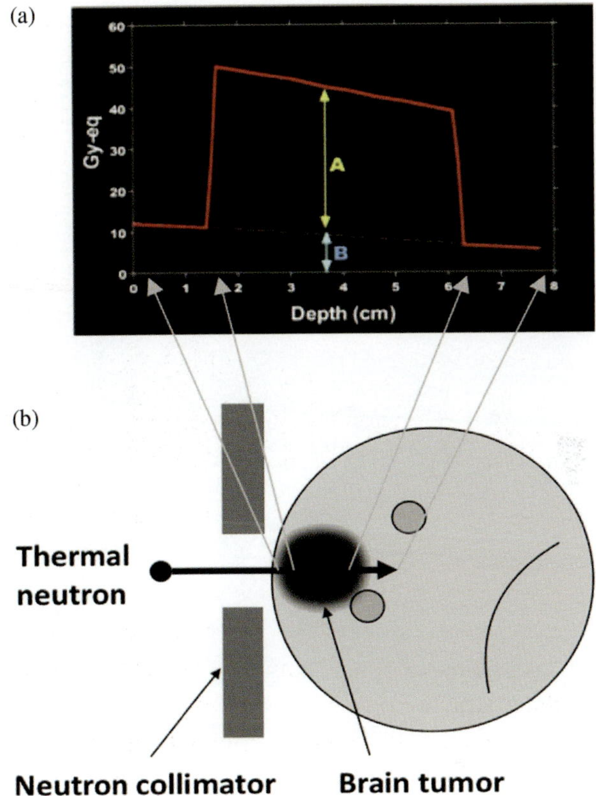

Figure 6.4 Dose distribution of BNCT for brain tumor (A: Tumor absorbed dose of ^{10}B(n,a)^{7}Li reactions; B: Non-selective absorbed dose of ^{14}N(n,p)^{14}C and γ-rays).

is well visualized as a high density area (Fig. 6.5 (a)).[316] Macroscopically, the boron distribution looks uniform and the tumor seems to be easily destroyed via thermal neutron bombardment. However, microscopically the boron distribution is very heterogeneous, especially in the marginal region of the tumor and also in the tumor due to the presence of quiescent cells (resting cell on G0 mitotic stage) that have only weak essential metabolic activity. Furthermore, the thermal neutron distribution is also heterogeneous, since the thermal neutron fluence strongly decreases with depth in deep brain, by reacting and scattering with nitrogen and hydrogen atoms in the brain parenchyma. Thus, after BNCT, the tumor was partially destroyed

(a)

(b)

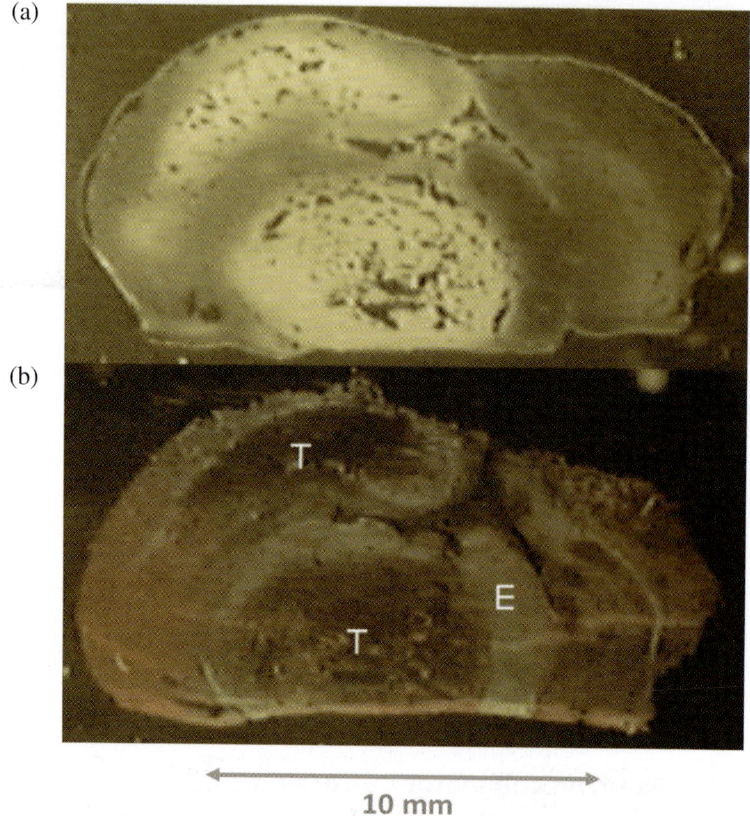

Figure 6.5   Alpha (α) autoradiography of boron in rat brain tumor (a): Alpha (α) auto-radiograph, (b) brain section tumor (**T**) and perifocal edema (**E**).

and is a mixture of lethal, sublethal, and intact tumor cells. In other words, "the tumor was overall near-killed."

As pointed out in an earlier chapter, with the initial use of slightly higher energy (1–10 keV) epithermal neutrons, the therapeutically effective thermal neutron distribution shifts in depth to around 2 cm, that is almost equal to the total thickness of scalp and skull as shown in Fig. 6.6. Thus, nonsurgical BNCT becomes practical for brain tumors.

Figure 6.7 shows the actual dose distribution of a case of BNCT without craniotomy. The absorbed dose reaches a maximum at a depth of around 2 cm from the scalp.

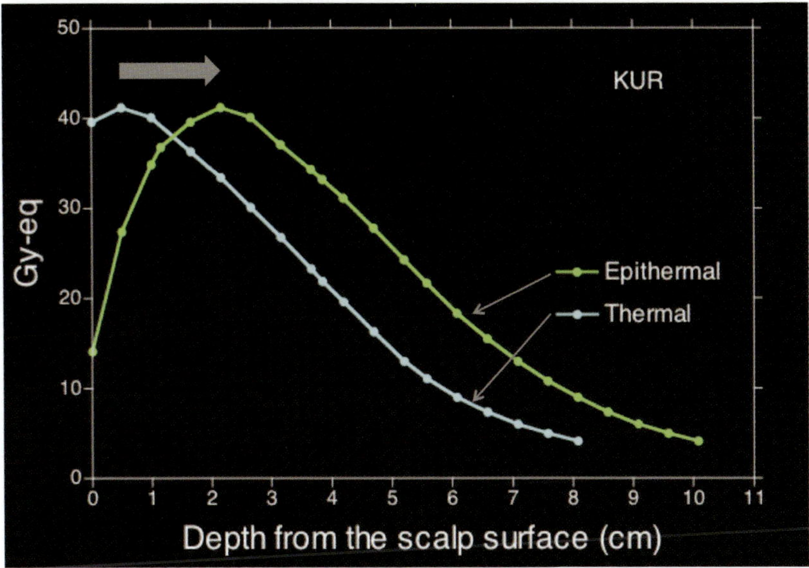

Figure 6.6 Thermal neutron distribution in tissue.

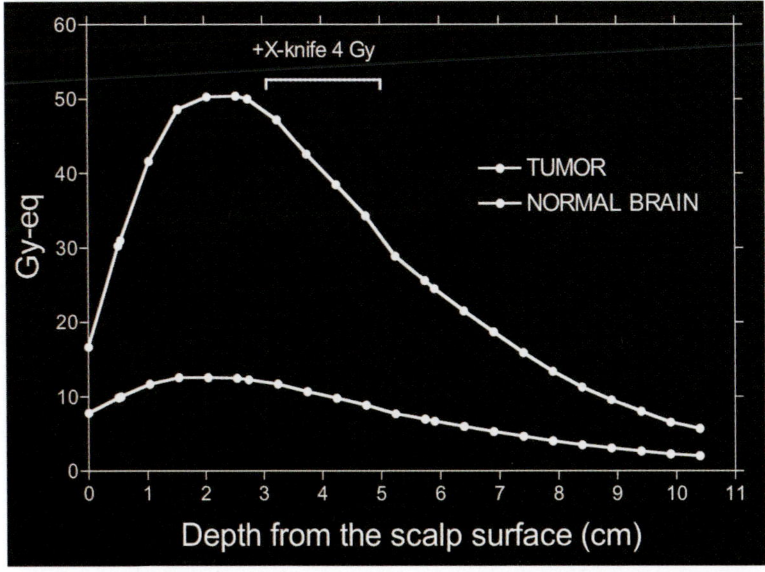

Figure 6.7 Absorbed dose distribution in brain tumor.

6.2. GdNCT Dosimetry

The radio-biological mechanisms of GdNCT and BNCT are quite different. In BNCT, heavy α particles and their recoil ^{7}Li ions transfer their total kinetic energy to the cells within which the capture reaction takes place, sparing other neighboring cells. Therefore, the microdistribution of kinetic energy in tissues is uneven. By contrast, the mixture of low- and high-energy ionizing particles used in GdNCT travel farther and their kinetic energies are more uniformly distributed throughout tumor tissues, except for Auger electrons (Fig. 6.8).

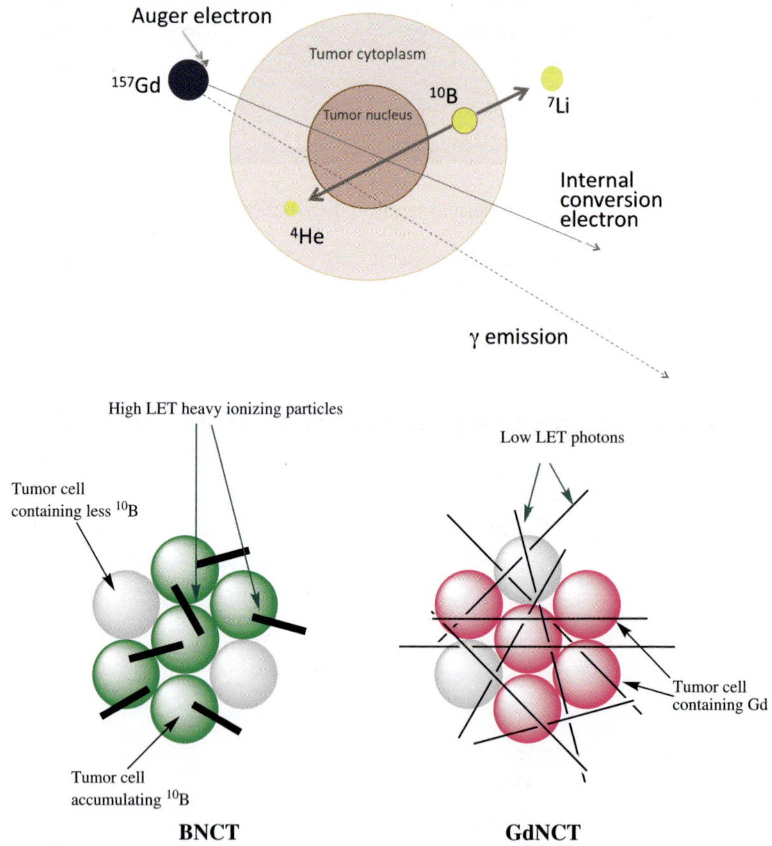

Figure 6.8 Schematic illustration of radiobiological differences between GdNCT and BNCT.

The different physical characteristics of GdNCT and BNCT are shown in Table 6.1. In BNCT, while the high-LET-heavy particles that are released are capable of killing tumor cells, the microdistribution of the radiation-absorbed dose is uneven due to heterogeneous boron distribution in tumor tissues, and highly localized energy transfer from the particles occur within ranges that do not exceed the dimensions of tumor cells. By contrast, while low-LET-ionizing particles are emitted in GdNCT, the absorbed dose distribution is more uniform. Thus, we should be able to use GdNCT to resolve some of the shortcomings of heterogeneous dose distribution found in BNCT.

The vascular endothelium is highly vulnerable to radiation, which causes vascular necrosis. While necrosis of tumor tissues at the brain surface is therapeutically desirable, doses received by vascular tissues should, in principle, be kept within tolerable limits. In GdNCT, the vascular dose is negligibly small because the low LET products causes little radiation damage (Fig. 6.9). There is also very little possibility of endothelial damage from Auger electrons generated from intravascular GdNCR (Gd neutron capture reactions). Indeed preliminary studies revealed no vascular necrosis from GdNCT after continuous infusion of high concentrations of GdDTPA.

Table 6.1 Difference in physical characteristics of GdNCT and BNCT.

	$^{157}Gd(n,\gamma)^{158}Gd$	$^{10}B(n,\alpha)^{7}Li$
Cross section	250,000 b	3985 b
Total kinetic energy	7.98 MeV	3.33 MeV
Heavy ion	—	High Let α and Li particle (RBE = 2–3, 14 μ)
Auger electron	High LET (RBE = unknown, several 100 ?)	—
High energy electron	Internal convergent electron (RBE = 1, several 100 μ)	—
High energy photon	Prompt γ emission (RBE = 1, several 100 cm)	—

Note: RBE: relative biological effectiveness, compares with standard (250 keV X-ray) radiation. Numerical values in parentheses indicate RBE values and trahectories of radiation particles.

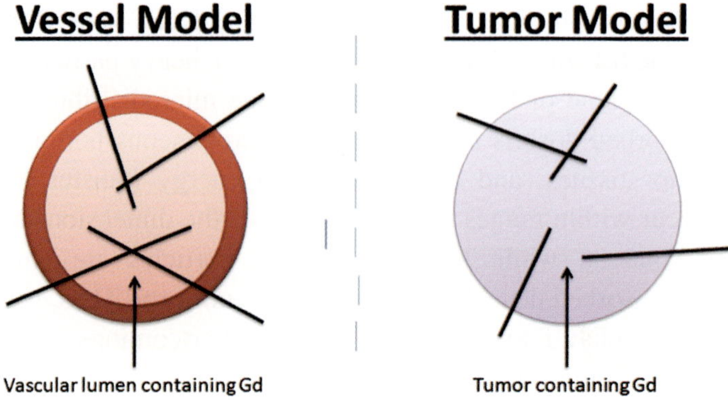

Figure 6.9 Different radiation dose distributions in blood vessels and tumors.

Why is vascular damage avoided in GdNCT? Because of the generally longer trajectories of the radiation (i.e. γ-rays, electrons, and X-rays) released by GdNCT, the larger the tumor, the greater is the general density of the absorbed dose within the tumor. Thus, since the radiation dose is directly dependent on tumor size, the tumoricidal effect of GdNCT is directly proportional to the size of tumor (bulk effect).

Gadolinium has an extremely large nuclear cross section for thermal-neutron capture. Since tissues containing Gd avidly absorb thermal neutrons, the presence of Gd in the upper levels of tissue tends to shield the lower levels from neutron bombardment. Consequently, the density of neutron capture decreases depending on tissue-transit depth. As a result, the number of thermal neutrons able to reach deep tumors (D) is inevitably insufficient for a therapeutic dose (Fig. 6.10). To minimize this shielding effect, the optimum ^{157}Gd concentration in tumors should be around 200 ppm ^{157}Gd (<1000 ppm ^{n}Gd).

In GdNCT, the dose distribution extends slightly beyond the tumor mass, and radiation injury may occur in the peri-region of the tumor infiltration (C) (Fig. 6.11). However, this radiation halo is not strongly tumoricidal beyond 100 mm from the edge of the main tumor.[317]

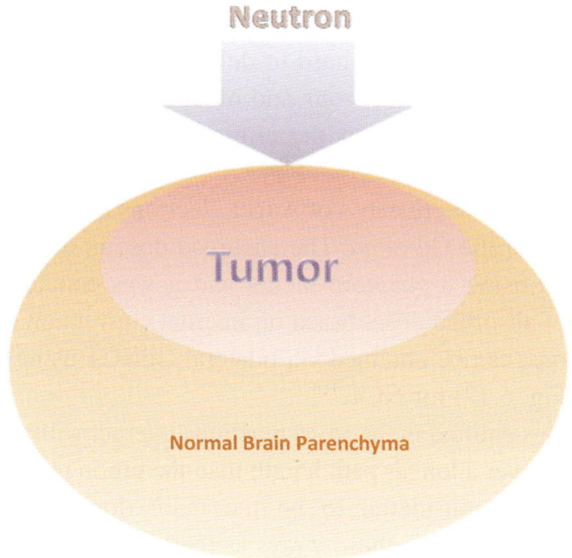

Figure 6.10   Self-shielding by ^{157}Gd.

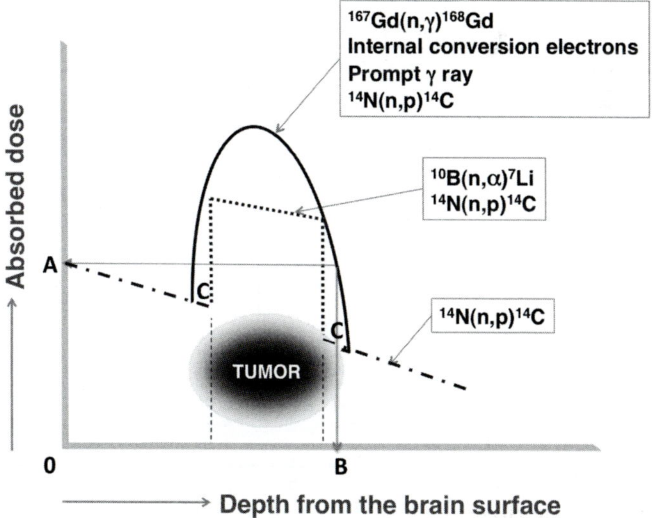

Figure 6.11   Limits of therapeutic dosage. **A**: Maximum tolerable dose to brain surface. **B**: Maximum therapeutic depth. **C**: The peri-region of tumor infiltration.

The thickness of the halo (100 mm) is equivalent to a thickness of only about seven tumor cells. It is hard to deliver boron compounds to this marginal area. Of course, neurons and other normal brain parenchyma are present in this region, but the clinical implications are probably minimal.

The biological effectiveness of Auger electrons is strongly dependent on the location of the Gd atom. The absorbed dose in deep brain tumors was calculated using a phantom model and it was concluded that very sharp peaks of absorbed dose, based on internal conversion electrons and Auger electrons, can be obtained via bilateral GdNCT using a D-T-based accelerator (Fig. 6.12) for NCT.[318]

The photons emitted in the (n, γ) reactions interact with the tissues but deposit energy over a longer path length than the boron reaction products. These γ-rays are considered to be the main drawback of GdNCT. However, if we can utilize these γ-rays as therapeutic in an optimal concentration of Gd, they may give sublethal damage to invading tumor cells in the surrounding tumor rim.

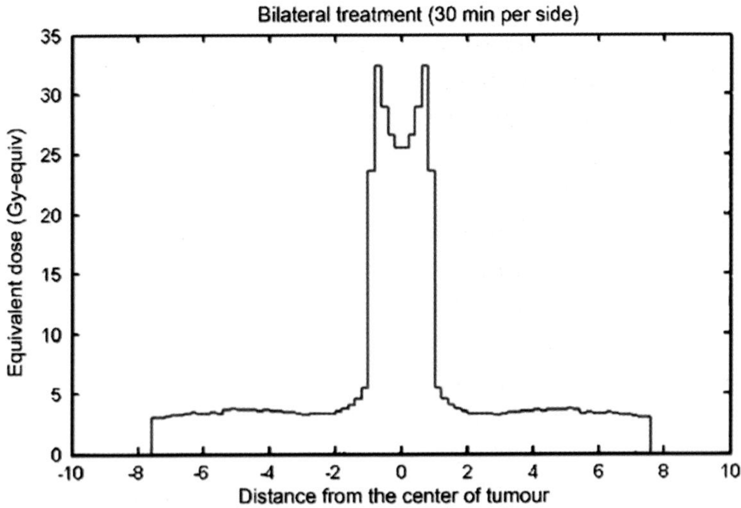

Figure 6.12   Equivalent dose in the case of bilateral treatment of 2 cm diameter tumor positioned at the center of the brain phantom containing 700 ppm ^{157}Gd in tumor and 5 ppm ^{157}Gd in normal brain.[318]

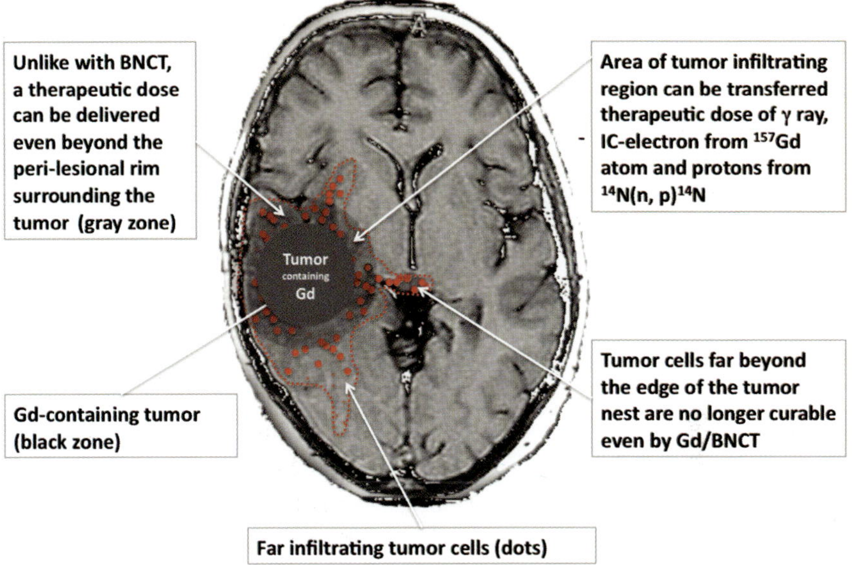

Figure 6.13 Illustration of a typical malignant brain tumor. Red spots indicate infiltrating tumors.

Currently, the strategy for treating malignant brain tumors that neuro-surgeons commonly employ is a maximum surgical extirpation of the tumor followed by adjuvant therapies such as radiation and chemotherapy. In this regard, post-operative adjunctive, fractionated GdNCT, after tumor extirpation, would be of limited benefit. However, GdNCT may provide a highly acceptable, noninvasive alternative treatment for malignant brain tumors. If a tumor malignancy can be determined by radio-pathological methods without any surgical exploration, fractionated GdNCT may become a highly reliable and acceptable treatment option for malignant brain tumors.

In GdNCT, tumor cells infiltrating close to the main tumor mass are susceptible to peripheral radiation from the main tumor area (Fig. 6.13). However, Gd/BNCT cannot treat tumor cells that have infiltrated far from the main tumor. In particular, tumor cells infiltrating to the contralateral hemisphere are beyond the scope of Gd/BNCT. Effective elimination of distantly infiltrating tumor cells, the prime cause of recurrence, remains a major challenge.

Chapter

Selected *in vitro* and *in vivo* Studies

7.1. *In vivo* and *in vitro* Studies Involving BNCT

There have been *in vitro* and *in vivo* studies in BNCT for more than 70 years, starting with the work of Kruger, Hall, and Goldhaber at the University of Illinois and at the University of California at Berkley.[69] Earlier chapters in this book, especially Chapters 3 and 4, discuss a number of potential boron delivery agents, along with the pertinent *in vitro* / *in vivo* studies. In addition there are a number of excellent reviews on this topic.[381,382] Since these discussions are comprehensive, there will be no attempt to repeat them here, instead several topics will be selected and discussed.

One of the main problems with most BNCT agents for the treatment of brain tumors is the difficulty in finding carriers that can cross the blood-brain-barrier (BBB). Of the two boron agents approved for clinical studies, only BPA can cross the BBB. If BNCT for brain tumors is going to advance, new compounds or delivery methods must be developed.

Among the techniques that should be more completely investigated is the use of nanoparticles as vehicles for transport across the BBB. Some of the most studied nanoparticles are liposomes. There are two approaches in loading the liposome with the ^{10}B containing compounds. One is to encapsulate the boron compounds in the liposomes; the other is to incorporate boron-containing entities in the lipid bilayer. As discussed earlier, one of the problems associated with the use of liposomes as drug delivery

125

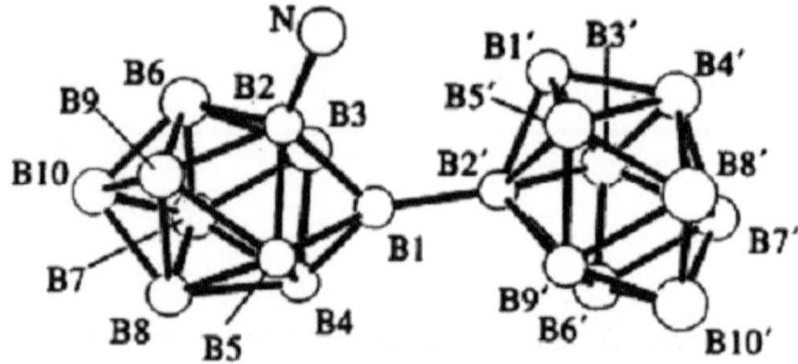

Figure 7.1   Crystal structure showing the heavy atoms of $1\text{-}(2'\text{-}B_{10}H_9)\text{-}2\text{-}NH_3B_{10}H_h]^{3-}$.[173c]

vehicles is the rapid rate of clearance by the reticuloendothelial system (RES). This decreases the circulation time of the liposomes and hence the extent of their localization in tumors. One way around this is to conjugate the liposome to PEG polymer units, to give so-called stealth liposomes that can avoid detection by the RES and exhibit extended circulation times. A PEG-conjugated liposome was prepared from distearoyl phosphatidylcholine (DSPC), PEG 2000-conjugated distearoyl phosphatidylethonalamine (PEG-DSPE) and cholesterol in a DSPC/cholesterol/PEG-DSPE ratio of 1:1:0.1, and charged with $[1\text{-}(2'\text{-}B_{10}H_9)\text{-}2\text{-}NH_3B_{10}H_h]^{3-}$ (crystal structure shown in Fig. 7.1).

When injected into BALB/c mice bearing EMT6 tumors, this liposome showed a continued accumulation in the tumor over an entire 48 hr experiment, reaching a maximum of 47 µg of B/g of tumor.[173c] *In vitro* studies, using cultured human KB squamous epithelial cancer cells, of a folate receptor (FR) targeted liposomes charged with $Na_2[BSH]$ and the $Na_3(B_{20}H_{17}NH_3)$ showed a maximum boron concentration of ~240 µg $B/10^9$ cells. The incorporation of folate-PEG-DSPE had no effect of liposome loading.[176] Epidermal growth factor receptors (EGFR) were conjugated to PEGylated ligand liposomes delivery vehicles loaded with water soluble boronated phenanthridine (WSP1) or water soluble boronated acridine (WSA1), shown in Fig. 7.2, and studied by Kullberg *et al.*[319] In the case of WSA1 in *in vitro* studies using glioma cells, a ligand dependent uptake was observed and the boron uptake was as good as for free WSA1; no ligand dependent uptake was seen for WSP1-containing liposomes. The results showed that 5.29 ± 1.07 µg B/g cell was obtained with WSA1.

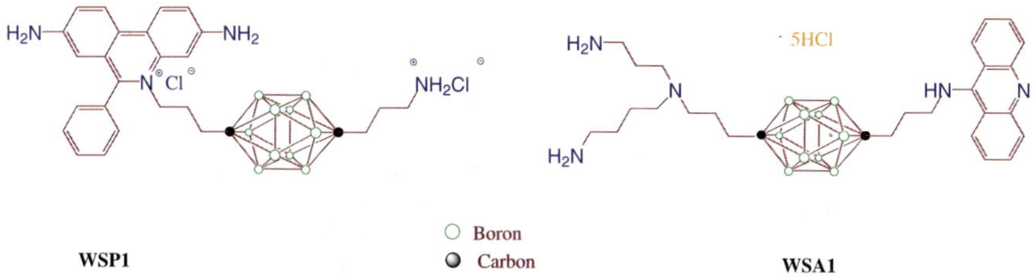

Figure 7.2 Chemical formulas of WSP1 and WSA1.

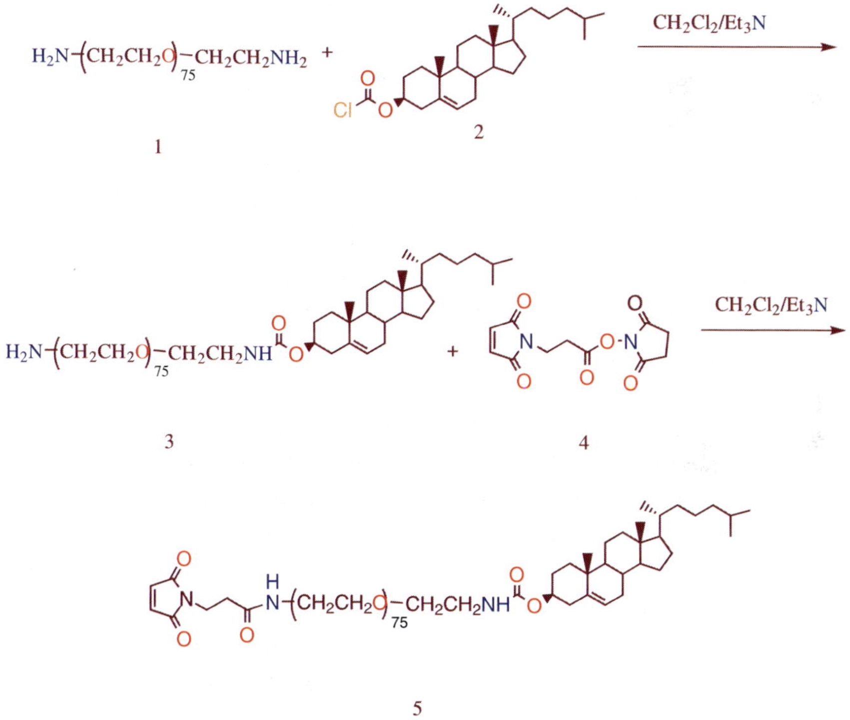

Figure 7.3 Synthesis of MAL-PEG-cholesterol.

Epidermal growth factor receptor (EGFR) targeting cetuximab-immuno-liposomes (IL) for targeted delivery of boron compounds to EFGR(+) glioma cells have also been studied.[320] The liposomes were obtained using a cholesterol-based anchor, malemido-PEG-cholesterol (Mal-PEG-chol), which was synthesized as shown in Fig. 7.3. The cetuximab monoclonal

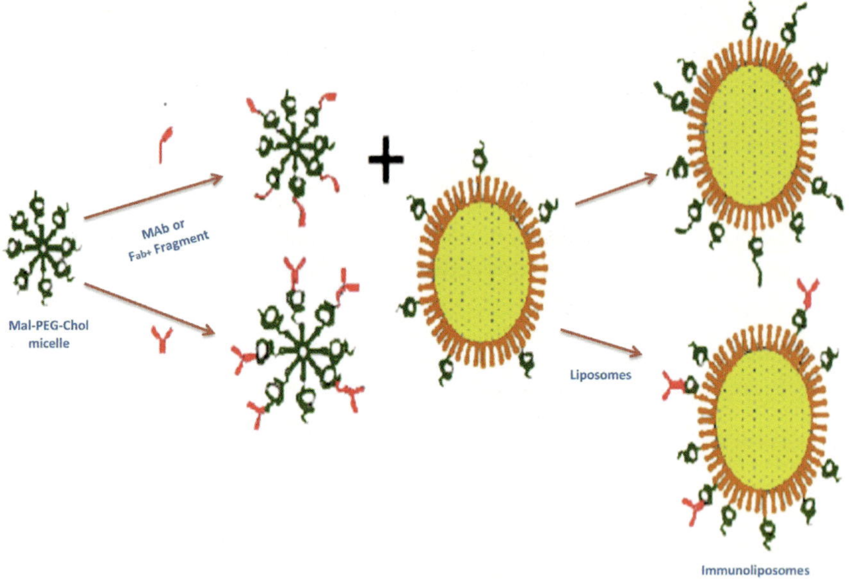

Figure 7.4   Synthesis of the cetuximab-immunoliposome from the Mal-PEG-chol micell and a preformed liposome.

antibody can be incorporated to obtain the cetuximab-immunoliposome as shown schematically in Fig. 7.4.

It was found that a ~8-fold cellular uptake of boron into EGFR(+) F98$_{EGFR}$ glioma cells compared to nontargeted human IgG-immunoliposomes.[320]

Given that immunoliposomes can be loaded with variety of polar and nonpolar molecules and used as delivery agents, some of which have been found to cross the BBB,[321] the future of BNCT for brain tumors could well revolve around the syntheses of these nanovehicles. There are a number of other liposomal systems for drug delivery; for further discussion, see the recent review by Nakamura.[323] There is also a recent discussion of the use of nanotechnology in BNCT.[322]

In addition to developing new, innovated delivery methods for approved boron drugs, new boron conjugates must be synthesized. Chapter 4 describes a number of boron-containing candidates that have been synthesized in response to this challenge.

Unfortunately, most of these new compounds have, at best, been studied only *in vitro* or *in vivo* using small animal (rat) models. In reviewing the literature on these compounds, the most encountered ending phrase is "this molecule warrants further study." That more times than not is the last sentence one reads on the subject, and the big three, BSH, BPA, and GB-10, stand alone in clinical trails. There are several systems that show promise and which could serve as a basis for extensive studies; they are the porphyrin analogues, including chlorin, bacteriochlorin, tetrabenzoporphyrin, corrole, and phthalocyanine macrocycles.[324] These compounds generally have high tumor selectivity and long tumor-retention times, low dark toxicities, and the ability to form complexes with DNA and RNA. Many of these compounds are used in photodynamic therapy (PDT).[325] Perhaps the two most-studied porphyrins are BOPP (40) and CuTCPH (41), shown in Fig. 7.5. Both were found to deliver high amounts of boron to tumor-bearing animals, with high tumor to blood and tumor to brain ratios. Phase 1 studies of PDT for high-grade gliomas using BOPP established a recommended dose of 4 mg/Kg of body mass.[326]

Cobaltacarboranes attached to a porphyrin core provides a way to increase the amount of boron and increase water solubility. Compound 30 (shown in Fig. 7.6) is an example of such a construct. The compound also

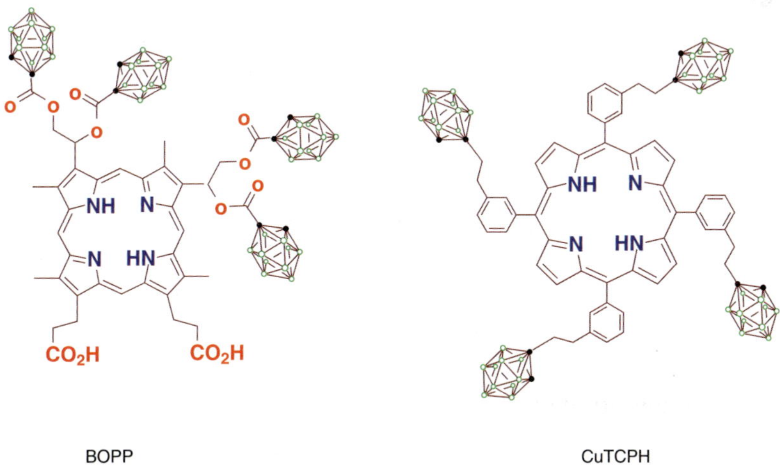

Figure 7.5   Boronated porphryins BOPP and CuTCPH.

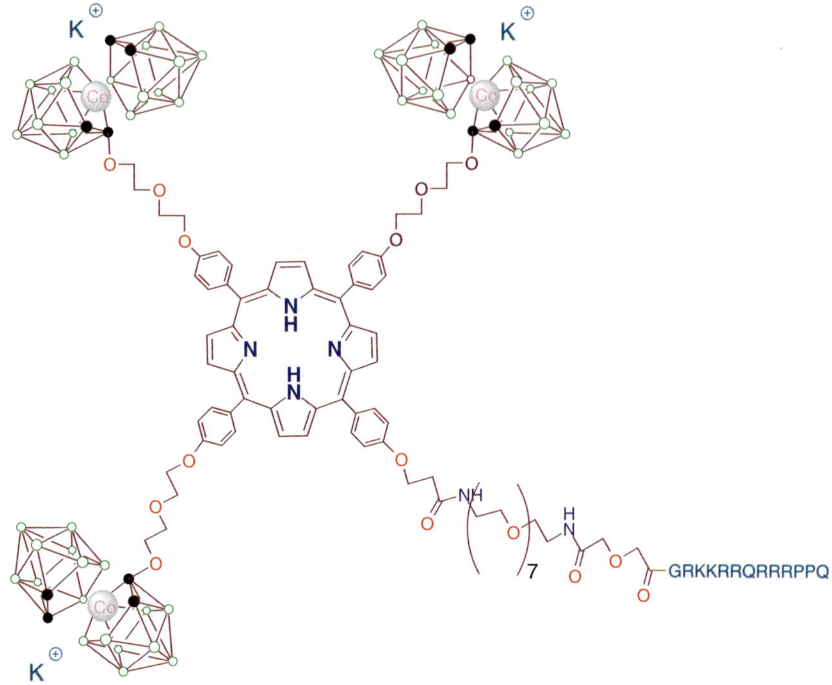

Figure 7.6   Cobaltacarborane porphyrin.

has a cell-penetrating peptide sequence, it was found that this molecule accumulated at higher concentrations in human Hep2 carcinoma cells than related molecules without a peptide sequence. The beauty of the porphryn-like derived molecules is that one has a powerful core around which different groups can be attached.

Another versatile group of biomolecules that are the building blocks of cells are the nucleosides. A number of boron-containing nucleosides have been synthesized and tested as potential boron delivery agents for BNCT. Boronated nucleosides could selectively accumulate in tumor cell due to their conversion to the corresponding monophosphates by cytosolic kinases (see Fig. 7.7). These are up-regulated in cancer cells.

Note that the NH group is not directly involved and the hydrogen can be substituted to give a series of boronated nucleosides, shown in Fig. 7.8.

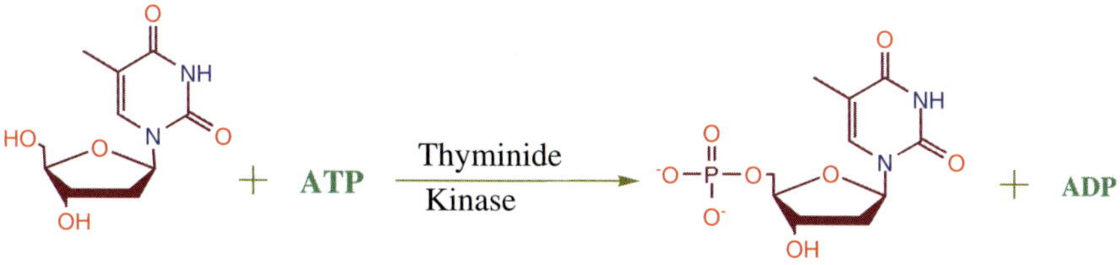

Figure 7.7 Catalyzed phosphorylation of thymidine.

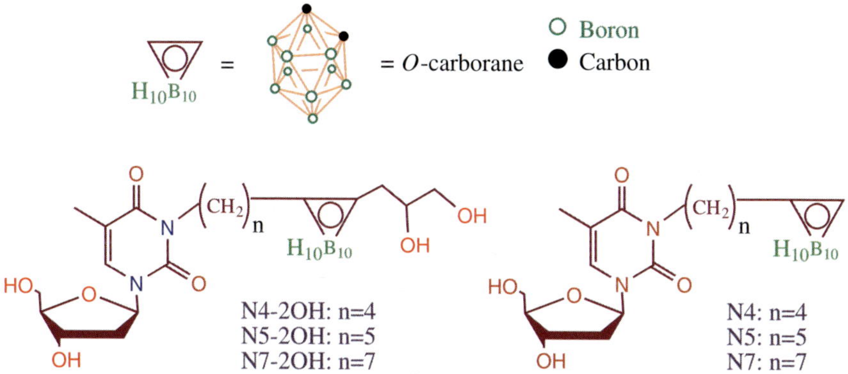

Figure 7.8 Structure and nomenclature of some 3-(carboranyl-alkyls and 3-(dihydroxypropylcarboranyl-alkyls thymidine (3CTA) analogs.

Their toxicity were tested against F98 rat glomia, human MRA melonoma, and murine L929 cell lines, all of which are thymidine kinase-1(+). The compound N5-2OH was found to be the least toxic and was selected for *in vivo* biodistribution studies in rats bearing F98 glioma and in mice bearing s.c. or intracerebral L929 tumors. At 2.5 hr after CED good tumor/brain ratios were found, indicating that there was selective retention by thymidine kinase-1(+) tumors.[327] Preliminary studies carried out in F98 glioma-bearing rats with ^{10}B enriched N5-2OH demonstrated a significant increase in survival after BNCT compared to the controls.[328] Studies were extended with the synthesis and testing of a library of 3CTA, as shown in Fig. 7.9.[329] All compounds were good substrates for recombinant human thymidine kinase-1 (TK1) with phosphorylation rates up to 89% relative to that of thymidine.[329] One

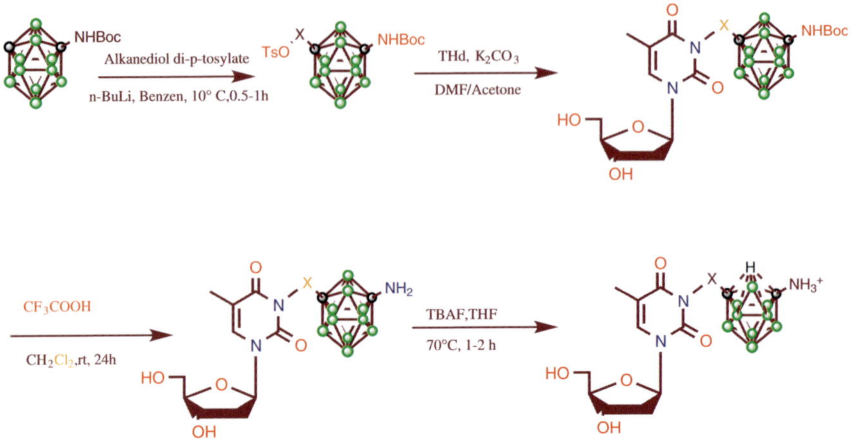

Figure 7.9　Syntheses and structures of some 3CTA compounds.[330]

compound, 19b in Fig. 7.9, showed selective retention in TK1-expressing murine L929 wild-type tumors versus L929 TK1(-) tumors in biodistribution studies.[329] Such systematic studies should allow the development of structure — activity relationships that could lead to more directed biosyntheses of compounds for BNCT. More studies of this type are needed.

7.2 *In vivo* and *in vitro* Studies Involving GdNCT

7.2.1 *In vitro studies involving GdNCT*

In 1989, Martin *et al.* reported that lethal damage due to double strand breakage (DSB) occurred using ^{157}Gd-DNA ligands by Auger electrons of gadolinium neutron capture therapy (GdNCT).[330] This was the first report of an *in vitro* study of GdNCT and it suggested the possibility of using GdNCT for malignancies. Akine *et al.*[331] reported that the tumor cell killing effect was 1.5 times greater than that of BNCT. The tumor cells were irradiated in *in vitro* condition at 5000 ppm nGdDTPA. They reported that 10% survival levels required 5.4×10^{12} neutrons/cm^2 without GdDTPA; on the other hand, a smaller dose of 1.55×10^{12} neutrons/cm^2 was required with 5000 ppm Gd (GdDTPA). They also demonstrated that

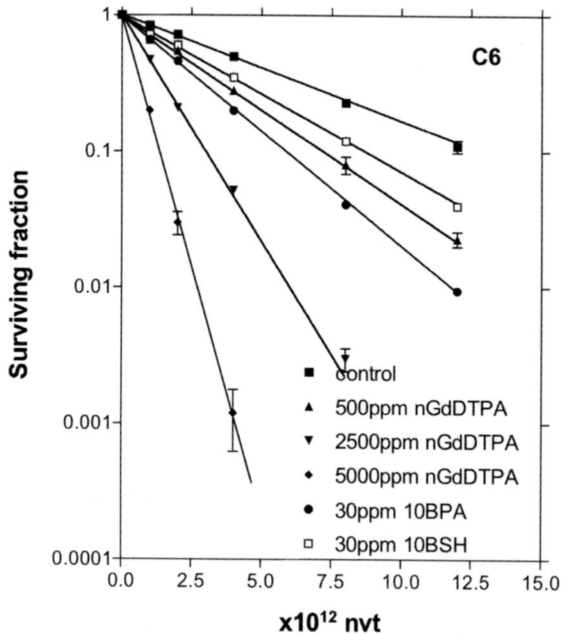

x10^{12} nvt

Figure 7.10   Surviving fractions after GdNCT using various Gd concentration, and comparative results for BNCT.

GdNCT by intraperitoneal injection of GdDTPA showed good tumor killing effect on Ehrlich ascites in *in vivo/in vitro* conditions.[331] These results suggested that GdNCT could be applied for malignancies using GdDTPA. However GdDTPA is not taken up into human tumor cells in *in vivo* condition, although it can distribute in tumor cells under *in vitro* conditions. This might be one of the pitfalls of *in vitro* GdNCT studies. Thus, it is essential to develop Gd compounds that accumulate in tumor cells with a long retention and that bind to DNA to establish GdNCT.

Figure 7.10 shows surviving fractions after *in vitro* GdNCT. Without a sigmoidal shoulder on the plot, survival decreased as a function of the thermal neutron dose: with 500 ppm Gd(+) medium, a 37% survival dose (D_{37}) was obtained with 3.55×10^{12} n/cm^2; with 2500 ppm Gd(+) medium, D_{37} was 1.40×10^{12} n/cm^2; and with Gd-free medium, D_{37} was 6.80×10^{12} n/cm^2. Compared with control survival, radiation with

GdNCT was 1.92 times more lethal with 500 ppm ^{n}Gd(+) and 4.86 times more lethal with 2500 ppm ^{n}Gd(+). Lethality with BPA at 30 ppm ^{10}B is similar to that of 500 ppm ^{n}Gd.

There are many reports that GdNCT might be promising as described in Chaps 4, 6, and 7. Especially GdNCT using liposome-encapsulating Gd ligands is a highly promising modality, the Gd concentrations in tumor is already at a clinically applicable level. GdNCT at a nontoxic concentration of gadolinium is effective in inducing cell death and chromosome aberrations in *in vitro* cell cultures.[334]

The *in vivo/in vitro* micronucleus (MN) assay is one of the standard methods to evaluate chromosome damage in early genetic toxicity investigations. Masunaga *et al.* reported that MN frequency was the largest in BSH, and then BPA, and the smallest in GdDTPA.[335] Their results suggest that Gd atom should be combined to DNA and/or critical organella in tumor cells for effective GdNCT.

Since thermal neutron capture cross section of Gd atom is very high, GdNCT might be possible even with the low fluence of thermal neutrons found in fast neutron therapy. Using the high energy accelerator of Fermilab, in the United States, a strong killing effect on tumor cells has been confirmed in *in vitro* GdNCT studies.[336]

7.2.2. *In vivo studies involving GdNCT*

Subcutaneous melanoma were investigated in larger tumor models. Ten days after subcutaneous implantation of 1 mm^3 of Green's melanotic melanoma into the thighs of Syrian golden hamster model (N = 30), the melanomas had grown to the size of a thumb tip ($\sim$12 $\times$ 12 $\times$ 12 $\div$ 3 mm^3 = 576 mm^3). For continuous infusion of GdDTPA, the femoral vein on the contralateral side was cannulated. During continuous 2 ml/kg/hr infusion of GdDTPA, thermal neutrons were irradiated on the thigh. During the protocols, the whole body of each animal was protected from neutron bombardment by a Li sheet (Fig. 7.11). After the start of infusion, the Gd concentration in blood was determined every 10 min by PGS. After GdNCT, tumor size was measured at intervals and tumor volume was estimated by dividing the result of multiplying the 3 measured dimensions by 3.

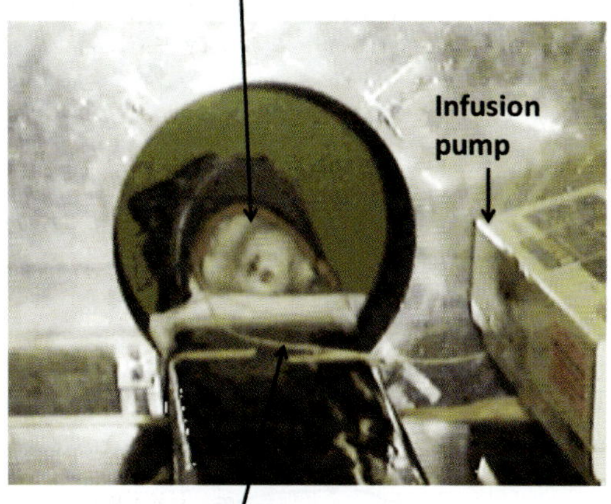

Figure 7.11 Hamster with cutaneous melanoma in thigh placed at the thermal neutron irradiation port of the Research Reactor Institute at Kyoto University.

As Fig. 7.12 shows, thermal neutron irradiation without Gd has little effect on tumor growth; by contrast, GdNCT was able to effectively destroy melanoma tissues. All tumor growth was strongly inhibited: tumor size dramatically decreased and was well restrained for more than two weeks after treatment. Complete destruction was also observed in 2 of 10 hamsters. We concluded that remission can be achieved by GdNCT with continuous infusion of GdDTPA.

Akine *et al.* reported similar results in that growth of gadolinium-infused tumors was significantly inhibited compared to that of control tumors ($p < 0.05$) between the 16th and 23rd days after treatment.[337] Intratumoral injection of liposome encapsulating Gd showed a long residence time in tumor cells, indicating that in addition to BNCT, liposomal delivery of Gd also might be promising for GdNCT.[338]

These results suggest that the gadolinium atom might be a promising nuclide for NCT when it accumulates in tumor. If the gadolinium compound does not bind to DNA, the Auger electrons cannot be effective. Mitin *et al.* have investigated the efficacy of antineoplastic action of GdNCT and BNCT in cases of canine melanoma and osteosarcoma.[339]

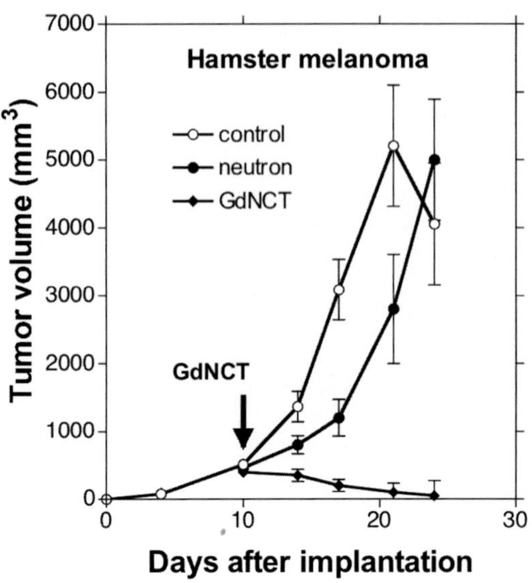

Figure 7.12   Effect of GdNCT on growth of subcutaneous melanoma in hamsters.

They compared 33 dogs with oral cavity melanoma and 9 dogs with osteosarcoma. They also concluded that GdNCT was a promising modality for NCT.

They cautiously noted that, in all cases of malignancies, both NCT types are methods of local tumor treatment, and distant metastasis can cause bad results of treatment, so the treatment should always be of a complex nature to minimize the risks of tumor spreading.[339]

7.3. *In vitro* and *in vivo* Studies Involving Combined BNCT and GdNCT

There are very few investigations concerning boron and gadolinium complexes (B-Gd complex). The purposes of such B-Gd complexes are as follows: (1) using Gd atom as a probe of boron on MRI (an MRI-BNCT agent), (2) using both therapeutic and diagnostic agent (B + Gd-NCT and MRI-BNCT). For example, carboranyl-DO3A (see Chap. 4.4) is a MRI-BNCT agent in which Gd atoms act as a proof by magnetic resonance imaging (MRI) for BNCT and are not therapeutic. Another Gd-carborane complex of C-(2-benzyloxy)-ethyl-C′-N-tert-butoxycarbonyl-aminomethyl-*o*-carborane is also a potentially useful intermediate for BNCT and MRI detector. This combination acts as a dual MRI-BNCT agent.[340]

On the other hand, BPA conjugated with GdDTPA can serve as B + Gd-NCT agent, where B and Gd are both therapeutic. Para-boronophenylalanine (BPA) conjugated with Gd-DTPA complex was reported to accumulate into tumor with a long retention time. Alpha-autoradiography revealed that it was accumulated in the tumors and the intestines, 20 min after injection.[341] Culbertson and Jevremovic showed theoretically that, depending upon the assumptions for drug specificity, tumor size, and fraction of quiescent cells, NCT with low levels of ^{157}Gd (125 ppm) supplementing ^{10}B loadings was superior to treatments applying ^{10}B alone.[342]

Chapter

Clinical Trials

8.1. Clinical Trials Using Boron Neutron Capture Therapy (BNCT)

This chapter will be divided into the various countries in which the trials are being conducted. Within each country heading, we will attempt to follow a chronological progression. However, since many trials went on simultaneously, a strict chronological schedule is not possible. Table 8.1 gives examples of clinical trials for the treatment of GBM by BNCT; GBM is by far the most investigated cancer in BNCT, its cure was the initial driving force for BNCT. It is only in recent years that clinical trials for other cancers have been initiated.

The studies will be subdivided into a brief description of early trials and a more extensive consideration of later trials. Early trials were all flawed by both lack of selective boron delivery agents and insufficient beam intensity. All these early trials failed and BNCT clinical trails in the United States were halted for over 30 years.

8.1.1. *Clinical trials in the United States of America*

8.1.1.1. *Early clinical trials in the USA*

The first BNCT trials in the United States were initiated in 1953 after the completion, in August 1950, of a nuclear reactor which was established for medical purposes at Brook Haven National Laboratory (BGRR).

Table 8.1   Recent clinical trials for GBM.

Facility	Dates	Number of Patients	Boron Drug	Infusion/Dose	Brain Dose Gy	Reference
HTR, MuITR, JRR, KURR Japan	1968–present	>250	BSH BPA	100 mg/kg 250 mg/kg	13 ^{10}B	289, 290
HFR Petten	1997–present	26	BSH	100 mg/kg/min	8–11.4 ^{10}B	291
BMRR BNL USA	1994–1999	53	BPA	250–330 mg/kg	20.3–36.0 (peak) 2.3–8.1 (average)	292–294
MITRR-II USA	1996–1999	20	BPA	250–350 mg/kg	8.7–15.4 (peak) 3.0–7.4 (average)	295
MITRR-II FCB MIT USA	2001–2003	6	BPA	350 mg/kg	NA	295
Studsvik AB Sweden	2001–2005	17	BPA	900 mg/kg (6hr)	7.3–15.5 (peak) 3.3–6.1 (average)	296
FIR1 Helsinki	1999–present	18	BPA	290–400 mg/kg	8.0–13.5 (peak) 3–6 (average)	297
LVR-15 Czech Republic	2001–present	5	BSH	100 mg/kg	<15.4 (peak) <2 (average)	298

Nineteen malignant brain tumors were treated at BGRR from 1953 to 1959. Seventeen cases of GBM were also treated at the research reactor of Massachusetts Institute of Technology in Boston during 1960–1961. The early clinical trials, conducted at BGRR and at the Massachusetts General Hospital (MGH) used thermal neutrons and sodium tetraborate (borax) as the ^{10}B vehicle. The patients had all undergone neurosurgical removal of malignant cerebral glioma but showed signs of recurrence. The irradiation time was from 17 to 40 min. There were no serious side effects from BNCT and the median survival time was 97 days. A second group of nine malignant glioma patients were treated with the less toxic sodium pentaborate with D-glucose in a 2:1 molar ratio and irradiated with a higher neutron flux and a higher ^{10}B concentration. Although the median survival time was 147 days, many patients developed severe, intractable radiation damage of the scalp.[70e]

One of the leaders in these early BGRR trials was Dr William H. Sweet, Professor of Neurosurgery at MGH. He subsequently carried out BNCT trials using the MITRR. However, because of insufficient neutron beam penetration, these BNCT trials demonstrated neither significant prolongation of survival nor evidence of therapeutic efficacy.[70] Because of the disappointing results, clinical trials in the United States were suspended.

8.1.1.2. *Later clinical trials in the USA*

Contemporary clinical experience with BCT started in 1994 with clinical trials at BNL and MIT.[349] These trials were characterized by the fact that epithermal neutron beams were used, ^{10}B distribution studies were undertaken and 3D Monte Carlo-based treatment plans were used. Twenty-two subjects were irradiated in these Phase I/II studies. The boron delivery agent in all cases was BPA-f, infused through central venous catheters at doses of 250 mg/kg over 1 hr (10 subjects), 300 mg/kg over 1.5 hr (2 subjects), and 350 mg/kg over 1.5 hr (10 subjects). The average age was 45 (24–78), and the study involved two subjects with metastatic melanoma and the remainder with gliobastoma multiforma (GBM). The tumor volumes varied in size, ranging from <10 cm^3 to 121 cm^3 as displayed by enhanced MRI. The most common side effect was alopecia (hair loss) and

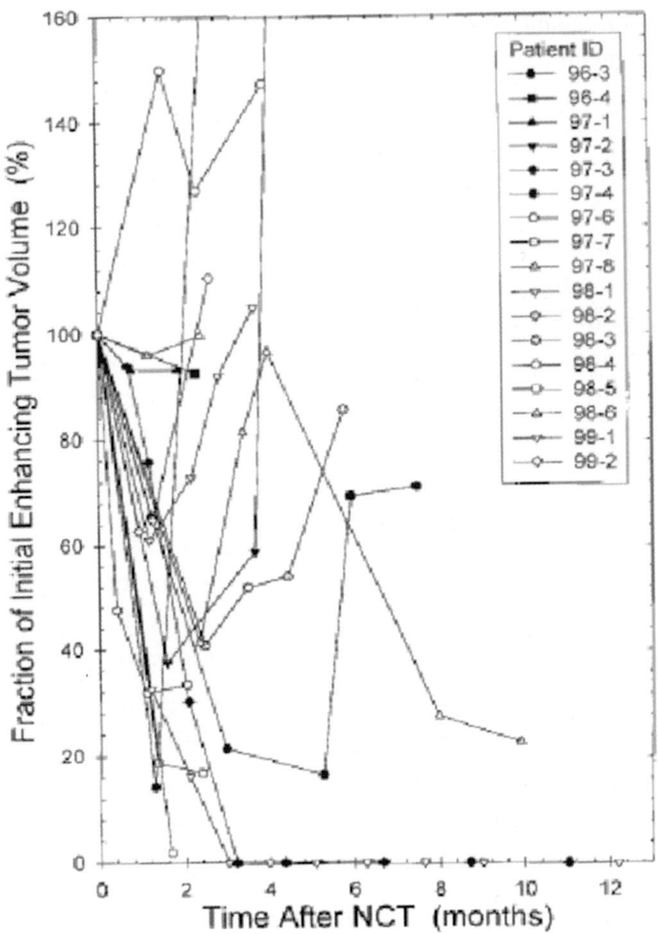

Figure 8.1   Contrast enhancement tumor volumes normalized to their original size.[349]

radiation dermatitis. As can be seen in Fig. 8.1, a majority (76%) experienced a sizable and progressive decrease in the tumor volume over the first several months, followed by a period of stability or regrowth.

The two subjects with metastatic melonoma showed a complete radiographic response. The median survival time for the subjects was 13 months, which was similar to that seen with resection and standard radiation. The studies did show that there were no major ill results due to the treatment.

8.1.2. *Clinical trials in Japan*

After early clinical trials were closed in the United States, BNCT research was largely transferred to Japan. Dr Hiroshi Hatanaka, Neurosurgeon of Tokyo University, who had studied BNCT with Professor Sweet, continued BNCT research with Professor Keiji Sano, Professor of Neurosurgery of Tokyo University in Japan; they treated the first patient with thermalized neutrons from the Hitachi Nuclear Reactor using BSH in 1968. Since that time over 183 patients with different kinds of brain tumors were treated by BNCT. There have been two reports by Nakagawa *et al.* on the results of these trials in 1997[73a] and 2003.[343] The earlier report covered the results from 149 patients (64, GBM; 39, anaplastic astrocytoma; 17, low grade astrocytoma; 30, other types of tumors) treated with BSH and thermal neutrons.[73a] The overall response rate in glioma patients was reported to be 64%. Seven glioblastoma patients (12%), 56% of the anaplastic astrocytoma patients, and 62% of the low grade astrocytoma patients lived more than two years. The MST of glioblastoma was 640 days (39–8138 days), MST of astrocytoma was 1811 days (17–6641 days), and of low grade astrocytoma was 1669 days (256–2638 days). Six patients (5 GBM, 1 anaplastic astrocytoma) died within 90 days of BNCT and six patients (2 GBM, 4 anaplastic astrocytomas) lived for more than 10 years. In assessing factors that might be related to patient survival, the patients were divided into two groups: group 1 being patients living less than two years after BNCT; group 2 being patients living more that two years after BNCT. Table 8.2 gives some possible prognostic factors for the two groups. As can be seen from the table, age seems to be statistically significant, target depth is important (group $1 = 6.1 \pm 3.5$ cm, group $2 = 4.8 \pm 3.2$ cm), as does neutron fluence at the target (group $1 = 3.8 \pm 4.1 \times 10^{12}$ n/cm^2, group $2 = 5.3 \pm 3.8 \times 10^{12}$ n/cm^2).

Hatanaka and Nakagawa reported possible long surviving patients of high grade glioma after BNCT for 120 cases. Out of 87 patients operated on before May 1987, in a report dated 1994, 18 lived or have lived or continued to live for longer than five years.[86]

However, Laramore pointed out that the median survival time of 12 high-grade glioma US patients who received BNCT in Japan during

Table 8.2   Prognostic factors.[74a]

	Group 1 (Less than Two Years)	Group 2 (More than Two Years)
Age	45.6 ± 14.1	37.6 ± 17.6
(years)	($n = 69$)	($n = 38$)
Radiation time	254.8 ± 106.1	249.6 ± 90.1
(minutes)	($n = 69$)	($n = 38$)
Target depth	6.1 ± 3.5	4.8 ± 3.2
(cm)	($n = 69$)	($n = 38$)
Neutron fluence	3.8 ± 4.1	5.3 ± 3.8
at target point	($n = 69$)	($n = 38$)
Maximum on the surface	14.1 ± 8.5	15.9 ± 8.1
of the brain	($n = 69$)	($n = 38$)
($E + 12n/cm^2$)		
Boron concentration		
in tumor	29.8 ± 17.3	21.7 ± 10.9
(ppm)	($n = 50$)	($n = 27$)
in blood	19.5 ± 14.5	15.4 ± 7.8
(ppm)	($n = 66$)	($n = 34$)
Physical dose at target point calculated by		
B-10 in tumor	$10.4 + 8.1$	11.3 ± 9.3
(Gy)	($n = 50$)	($n = 27$)
"26 ppm"	9.3 ± 5.2	11.7 ± 6.8
(Gy)	($n = 69$)	($n = 38$)
B-10 in blood	7.3 ± 6.0	6.6 ± 4.0
(Gy)	($n = 66$)	($n = 34$)

n = number of patients.

1987–1994 was 10.5 months and long survival cases were confined to anaplastic astrocytoma.[355]

Figure 8.2 shows the survival fraction of an early BNCT Phase II multicenter study at KURRI (Kyoto University, Kyoto Prefectural University of Medicine, Kagawa National Children's Hospital). This was not a randomized study; contralateral invasion and intraventricular dissemination were eliminated from this study. Gross five year survival for malignant brain tumors was around 20%, 50% for grade III glioma, such as anaplastic astrocytoma, and 10% for grade IV glioma such as glioblastoma.

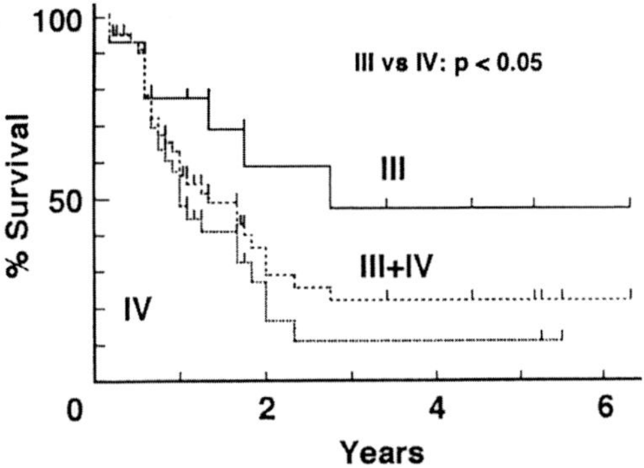

Figure 8.2   BNCT outcome of Phase II Multicenter study at KURRI.

A later report by Nakagawa *et al.* in 2003 was a retrospective study of 105 patients with gliomal tumors who were treated between 1978 and 1997.[343] Ten patients were treated with a mixed beam of thermal and epithermal neutrons. They reported a standard technique for the mixed beam therapy. First is to debulk the tumor and create a cavity during a preliminary operation performed about 1–2 weeks before BNCT. Approximately 12–13 hr before BNCT 100 mg/kg of BSH was delivered, then a thin silastic rubber balloon was inserted in the cavity thereby maintaining the size of the cavity during neutron irradiation. The minimum tumor volume dose was 15 Gy; the minimum target volume dose was 18 Gy; the maximum vascular dose did not exceed 15 Gy; and the total amount of gamma rays was below 10 Gy. Out of 10 glioblastoma patients treated, five died, in three of these, the cause was dissemination of tumor cells in the subaracnoid space and autopsy showed no recurrence around the primary tumor site, in the other two deceased patients, CT or MRI showed tumor recurrence. The remaining five patients were still alive at the time of the 2003 report.

A later report of the treatment of 21 cases of newly diagnosed malignant gliomas treated between 2002 and 2007, involved epithermal neutron irradiations and a simultaneous use of BPA and BSH.[353] Ten patients were treated with BNCT only and 11 with BNCT plus 20–30 Gy of fractionated external beam external X-ray irradiation. Treatments were well tolerated. The mean overall survival of patients treated with BNCT alone was

14.1 months, with BNCT plus X-ray, 23.5 months, compared to a historical control of 10.3 months.

Nakagawa *et al.* also reported on clinical trials of malignant brain tumors in children (under 15 years).[354] This was a subset of a larger 183 patient study.[343] The results indicate that instead of conventional radiation, BNCT can be applied to malignant brain tumors of children, especially those under three years.

An example of a typical case of early BNCT that showed intracranial metastasis after successful BNCT is that of a 5-year-old girl who complained of the onset of an acute headache and vomiting at her kindergarten on the Christmas Day. Computed tomography (CT) scan of the brain revealed a huge tumor mass with irregular margin and ring-like enhanced pattern in the right parietal lobe, thalamic region, as shown in Fig. 8.3 (left). Two weeks after onset, the patient received BNCT at Musashi Institute of Technology. In order to make thermal neutrons penetrate into deep brain, the patient received D_2O administration overnight and D_2O irrigation of ventricular system during BNCT. BNCT was very successful, the patient was discharged without neurological deficits two weeks after BNCT. Figure 8.3 (right) shows an enhanced CT scan two months after BNCT.

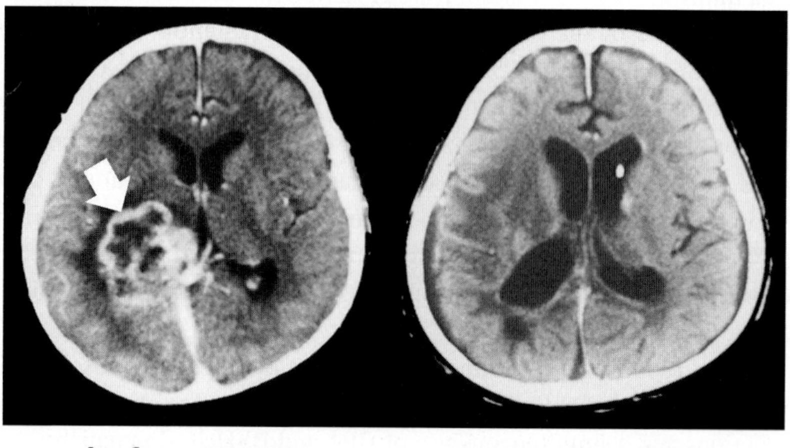

Figure 8.3 Computed tomography (CT) scan of the brain enhancing with a contrast agent.

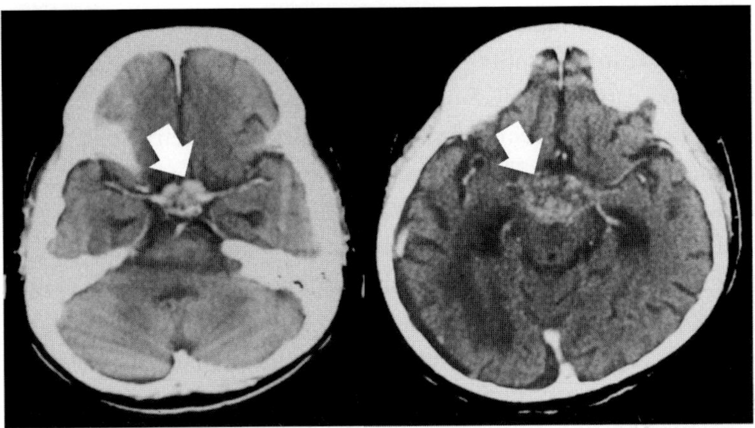

Figure 8.4 CT Scan of the brain 10 months after BNCT revealing an intracranial metastatic lesion in the basal cistern and hypothalamic region.

Unfortunately, 10 months after BNCT, the patient developed diabetes insipidus, a CT scan with enhancement of the brain revealed an intracranial metastatic lesion in the basal cistern and hypothalamic region (Fig. 8.4). There were no apparent lesional recurrence on the right parietal original region. The patient died from hormonal disturbance 13 months after BNCT.

This illustrates the weakness in current BNCT. The main tumor can be treated successfully by BNCT. However, tumor cells invading away from the main tumor cannot be controlled. The important strategy for malignant tumor treatment is not the control of the main tumor mass but of the invading tumor cells. This requires follow-up chemotherapy.

In very recent studies, the use of epithermal neutron-based BNCT, using a combination administration of BSH and BPA during nonsurgical BNCT, accompanied with a 30 Gy booster photon irradiation on the deepest lesion is proving to be a principal strategy for the current clinical Phase I / II BNCT studies for high grade gliomas. Chemotherapies, such as with temozolamide, are employed when tumor recurrence is suspected. In these studies, the median survival time (MST) was reported to be around 25 months; that is almost twice that found in early studies.[356–358] Although these data are not randomized, and the protocols are slightly different among the investigators, these are important results. With the development of new boron carriers, BNCT may prove to be a highly effective treatment.

BNCT has also been employed against recurrent head and neck malignancies (HNMs). The first report was by Kato *et al.* in 2004.[359] Six patients with HNM were treated with BNCT. The boron delivery agents were BPA and BSH and epithermal neutrons from the Kyoto University Research Reactor (KUR) were employed. The results showed: (1) the ^{10}B T/N ratios for SCC were 1.8–4.4, for sarcoma, 3.1–4.0, for parotid tumors, 3.5. (2) The relative volume of each tumor after treatment was 6–46% that of the original. (3) A remarkable reduction from 46–100% was observed for very large tumors (40–675 cm^3). (4) A marked improvement in the quality of life (QOL) and very mild side effects were observed. More recent studies by Kato *et al.* on advanced cases of HNMs show that more than 24% of the patients had a 6-year survival.[360] Kato *et al.* reported that survival periods after BNCT were 1–72 months (mean: 13.6 months) and the six-year survival rate was estimated to be 24% by a Kaplan–Meier analysis (see Fig. 8.5). Figure 8.6 shows the change in clinical appearance of a 67-year-old patient 22 months after the first BNCT treatment. The effectiveness of BNCT, as demonstrated by the remarkable reduction of the huge tumor, the disappearance of ulceration with normal skin cover, and the continuing improvement of facial palsy, testifies to the tumor-selective character of BNCT.

These results indicate that BNCT represents a new and very promising treatment for advanced HNM.

8.1.3. *Clinical trials in Finland*

In May, 1999, the first GBM patient was treated by BNCT using the epithermal beam of the Finnish Research Reactor (FiR1). Figure 8.7 shows a schematic of the treatment facility. Between May 1999 and December 2001, 18 patients with supratentorial glioblastoma were treated with BPA as the boron carrier.[362] All patients underwent prior surgery, but none received any chemotherapy or conventional radiotherapy. The patients were given 290–400 mg/kg of BPA-fructose prior to neutron beam irradiation. The average planning target volume dose was 30–61 Gy (W) and the

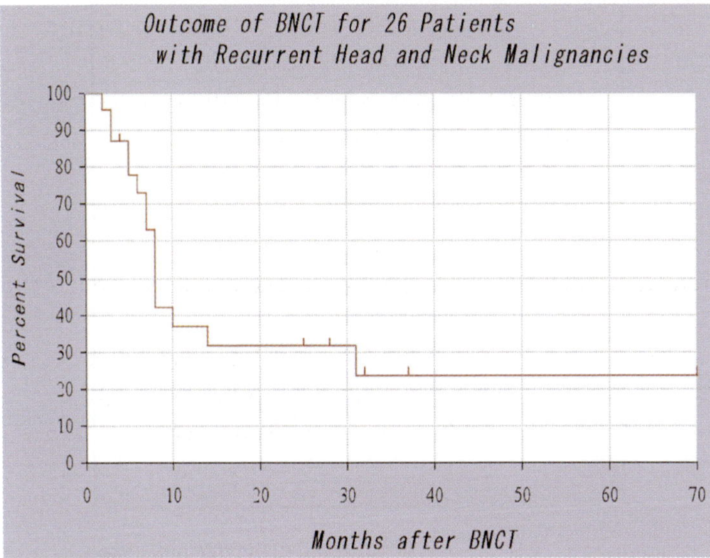

Figure 8.5 Kaplan–Meier analysis on 26 patients with recurrent head and neck malignancies.[361]

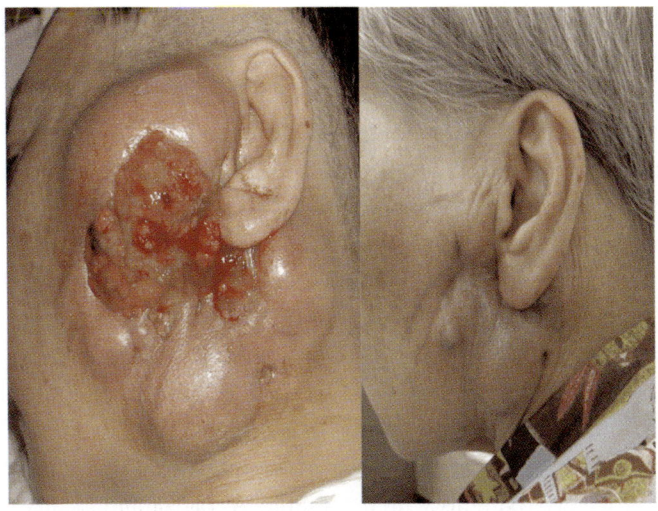

Figure 8.6 Clinical appearance of a 67-year-old patient with adenoid cystic carcinoma. Left: Before BNCT, Right: 22 months after BNCT.[361]

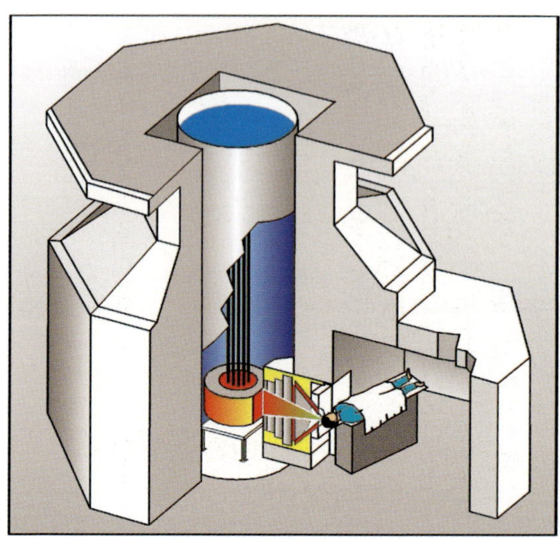

Figure 8.7   Schematic drawing of the BNCT facility at FiR1.[362]

average normal brain dose was 3–6 Gy(W); irradiation was given as a single fraction from two fields. The treatment was well tolerated; no patient deaths occurred during the first few months of treatment. The six-month overall survival rate was 100%, and the one-year overall survival was estimated to be 61%. Figure 8.8 shows the MRI results of BNCT.

Another trial, designed to see whether BPA-based BNCT is feasible in patients with recurrent or progressive glioblastoma who have received prior external beam radiotherapy was opened in February 2001, and as of 2003, three patients had been treated.[362] Therapy consists of surgical debulking, BPA-fructose at 290 mg/kg and irradiation given as a single fraction, through two portals. The brain peak dose was limited to less than 8 Gy(W), the average brain dose to ≤ 6 Gy(W), and the minimum planned tumor dose to ≥17 Gy (W). The protocol is planned for a total of 22 patients. Of the three completed, two died of progressing glioblastoma, 5 and 7 months after BNCT and one was alive after 12+ months.[362]

There have been excellent results in a clinical trial involving head and neck cancer. Twelve patients with inoperable, recurred, and locally advanced head and neck cancer were treated with BNCT in a single center Phase I/II clinical trial.[363] Prior treatments consisted of surgery, photon irradiation to a

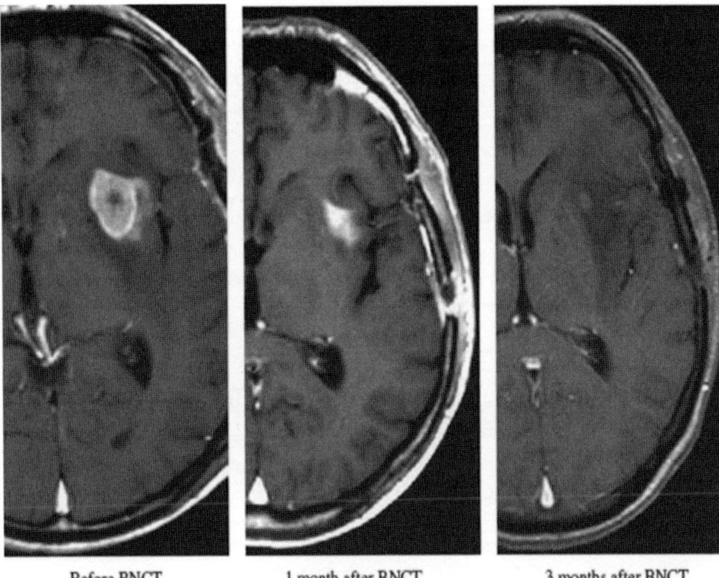

Figure 8.8 Radiation results in a 39-year-old man with GBM. Left panel: MRI scan taken 10-days after brain surgery, Middle panel: MRI taken 1-month after BNCT; Right panel: MRI three-months after BNCT.[362]

cumulated dose of 56–74 Gy. Each patient was infused with 400 mg/kg of BPA-fructose. Ten patients were irradiated twice and two were treated once. Ten (83%) responded to BNCT (seven complete response, three partial response), and the remaining two had tumor growth stabilization for 5.5 to 7.6 months. Overall the median time to disease progression was 9.8 months and the median survival time was 13.5 months. At 12.8 to 19.2 months after treatment with BNCT, five patients remained alive and four remained free of cancer. Therefore, BNCT has been proven to be a safe and effective treatment of inoperable, locally advanced head and neck tumors.

8.1.4. *Clinical trials in Sweden*

A comprehensive BNCT facility was constructed at the R2–00 research reactor in Studsvik, Sweden. Figure 8.9 shows a schematic of the beam configuration.[364]

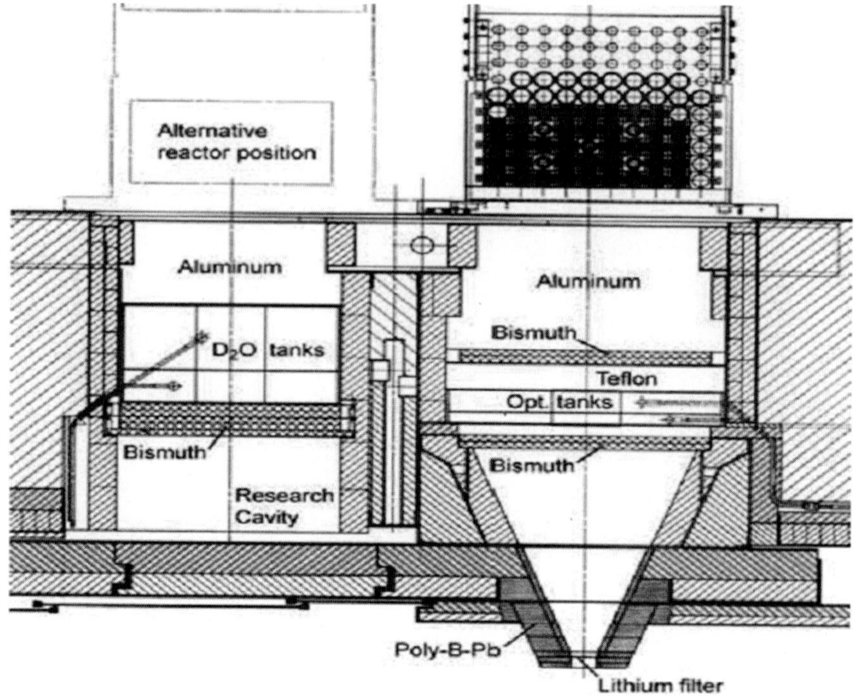

Figure 8.9   Horizontal cross section of the beam configuration at the Studsvik facility.[364]

A Phase II clinical trial was initiated on BNCT for GBM using BPA and epithermal neutrons.[365] One of the objectives of these trials was to investigate the effects of BNCT using a 6-hr infusion time of 900 mg/kg body weight of BPA-fructose, with neutron irradiation starting 2 hr after completion of the infusion. Thirty patients (26–69 years) with GBM were included in the trials; 27 had undergone debulking surgery. The average weighted dose to normal brain was 3.2–6.1 Gy(W), the minimum dose to tumor volume ranged from 15.4 to 54.3 Gy(W). The median time from BNCT to tumor progression was 5.8 months and the median survival time after BNCT was 14.2 months. Following progression, 13 patients were given temozolomide, patients treated with temozolomide lived considerably longer (17.7 versus 11.6 months). Figure 8.10 shows a

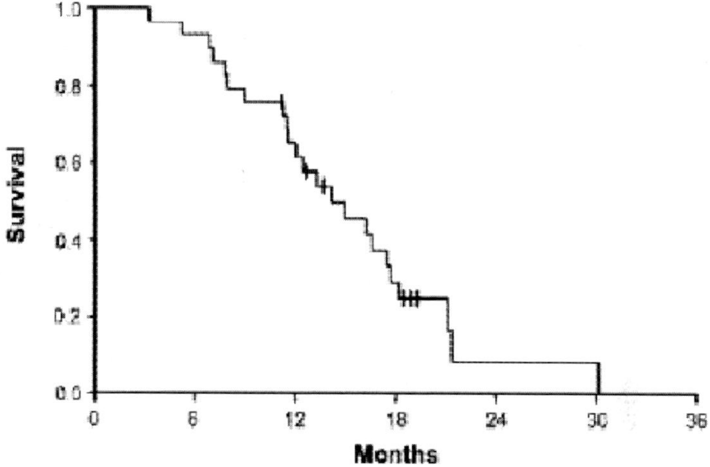

Figure 8.10 Kaplan–Meier overall survival curve after BNCT.[365]

Kaplan–Meier overall survival curve after BNCT. Seven patients suffered from seizures, eight from skin/mucous problems, five patients were stricken by thromboembolism, and four from abdominal disturbances in close relation to BNCT.[365]

The results to date show that BNCT is comparable to conventional forms of radiotherapy but the treatment time is much shorter. In view of the side effects in the present study, it probably will not be possible to increase the dose of BPA. More efficient boron carriers must be developed.

8.1.5. *Other clinical trials*

Most of the trials discussed above have involved the use of BNCT in combating brain cancer, specifically GBM. There have been additional case histories and trials on treatments of other types of cancer at various nuclear facilities. A Phase I/II melanoma clinical trial was conducted by the Argentine Atomic Energy Commission (CNEA) and the University of Buenos Aires Oncology Institute Angel H. Roffo (IOAHR).[366] Between October 2003 and June 2007, seven patients received eight treatment sessions covering 10 anatomical areas located in their extremities. All patients had multiple subcutaneous skin metastases of melanoma. Each patient received an infusion containing $14g/m^2$ of ^{10}BPA and followed by exposure of the particular area by a mixed thermal-epithermal neutron beam. The maximum prescribed dose to normal skin ranged from 16.5–24 Gy-Eq and normal tissue administered dose varied from 15.8–27.5 Gy-Eq. An overall response was observed in 69.3% of the nodules defined as a target, with the remainder showing no change. No progression of disease was observed inside the treatment boundaries on follow up. Figure 8.11 shows the results of BNCT (a) before, (b) one month after and (c) six months after BNCT.[366] As a result of the clinical observations, 24 Gy-Eq was considered the prescription dose for normal skin.

There have been several reports on the treatment of liver cancer with BNCT. One of the most innovative treatment procedures was that reported by Zonta *et al.* in Pavia, Italy[367] (see Fig. 8.12). The procedure involved three phases: a surgical phase in which the liver, infused with ^{10}B is removed; an irradiation phase in which the liver is irradiated with thermal neutrons; second surgical phase in which the liver is reconnected to the patient. The procedure was first preformed in 2001 on a 48-year-old male patient with diffuse metastases of a colon carcinoma. The patient was infused with 300 mg/kg body weight of ^{10}BPA, then under went a hepatectomy, the isolated liver was washed, chilled, and transferred to the thermal chamber of the Triga Mark II nuclear reactor and irradiated with a neutron fluence of 4×10^{12} n/cm^2; biopsies of metastatic and normal hepatic tissue showed a ^{10}B concentration ratio of about 6:1. After irradiation, the liver was reimplanted. The first patient so treated survived for 44 months with a good quality of life but died because of recurrences of his intestinal tumor (see Table 8.3 and Fig. 8.13).[367] In 2003, a second patient underwent the same procedure, but after 33 days died of cardiac failure. This patient suffered from dilative cardiomyopathy (see Table 8.4 and Fig. 8.14).[367]

This clinical experience, though limited, does demonstrate this could be an acceptable therapy, provided the patient can withstand such traumatic operations. Wittig *et al.* have described a trial (EORTC 11001) to determine whether BPA or BSH are viable candidates as boron carriers in such a procedure.[368] In this trial, patients who were about to undergo resection of hepatic metastases were infused with ^{10}BPA,

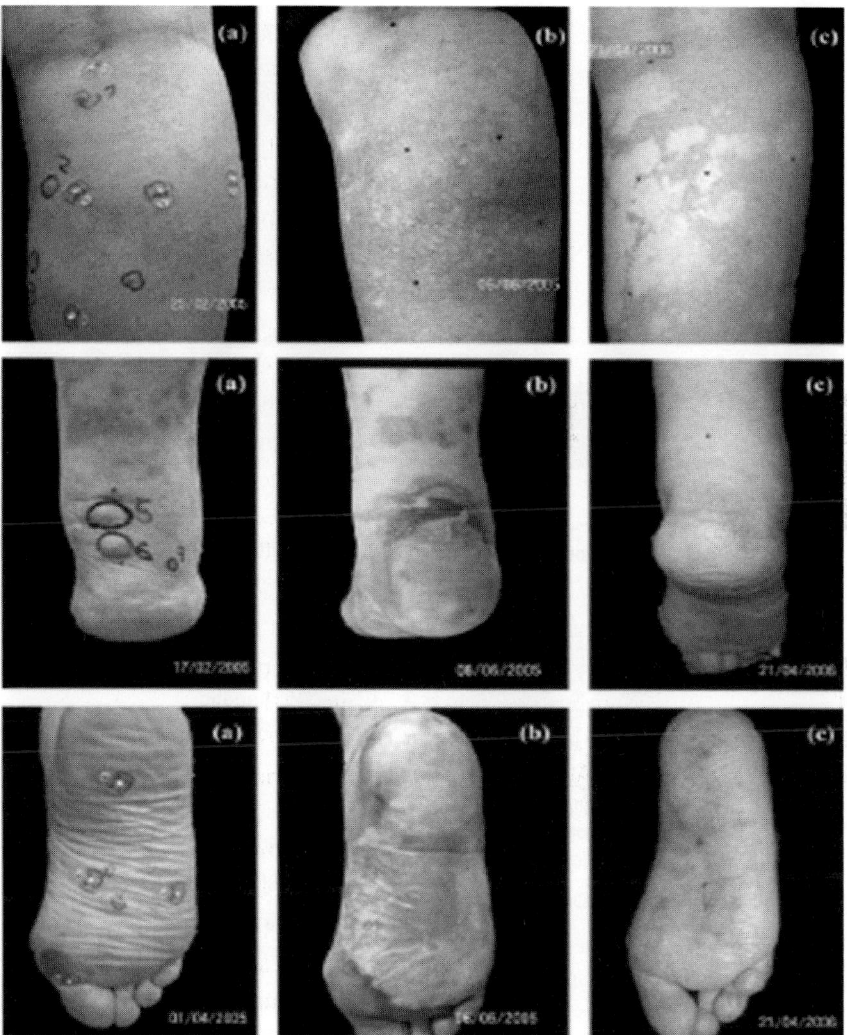

Figure 8.11 Effects of BNCT. (a) Before BNCT (b) one month after BNCT (c) six months after BNCT.[366]

^{10}BSH, or combinations thereof, and samples of blood, liver metastases, and normal liver were taken and analyzed for ^{10}B by prompt gamma ray spectroscopy (PGRS). Three patients were infused with BPA, three with BSH, and three with a combination of BPA and BSH. For BSH, the

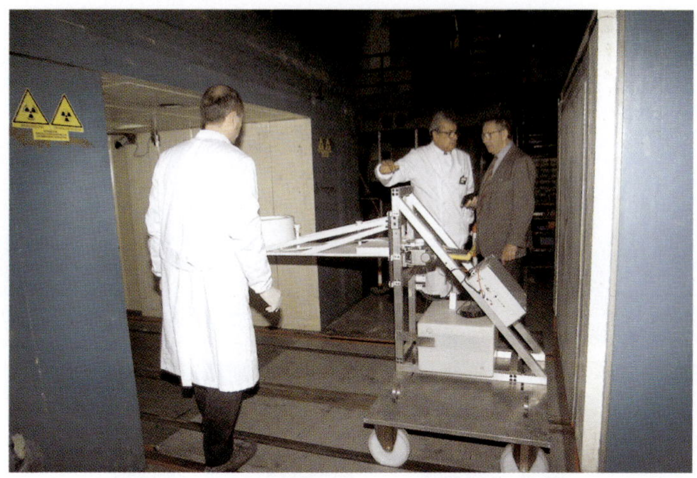

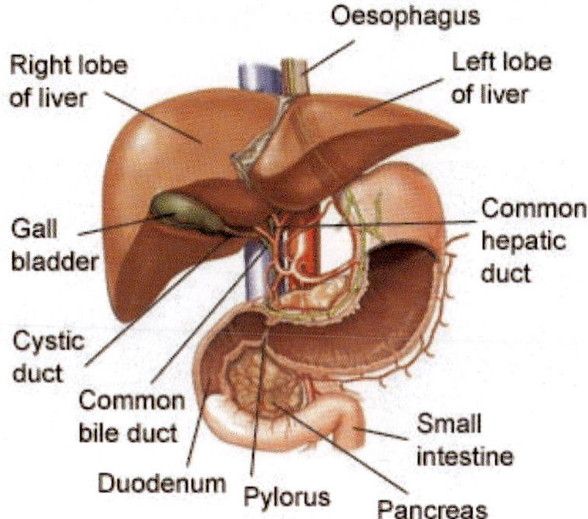

Figure 8.12 Liver explantation at the Taormina BNCT project. *Above*: BNCT preparation; *Below*: A perspective view of the human liver.[367]

highest ^{10}B concentration was found in the liver (31.5 ± 2.7 µg/g), followed by blood (24.8 ± 4.7µg/g), and then tumor (23.2 ± 2.1 µg/g). For BPA, the highest ^{10}B concentrations were found in the tumor (12.1 ± 2.2 µg/g), followed by the liver (8.5 ± 0.5 µg/g), and then blood (5.8 ± 0.8 µg/g). These results show that BSH is not a suitable boron carrier for liver metastases; however, BPA accumulates in the liver tumors to a sufficient extent that it could be a viable boron carrier for the extracorporal irradiation of liver in BNCT.[368]

Table 8.3 BNCT Pavia project. From the clinical record of the first patient.[367]

Sex and age:	Male, 48-year-old
Primary tumor:	Sigmoid adenocarcinoma p T3 G2 N1 M1[a]
Vital organ state:	Good
Hepatic involvement:	14 small bilobar synchronous metastases
Residual liver function[b]:	63%[c]
Vascular abnormalities	None
Dose of 10 BPA	300 mg/kg body weight through a colic vein
Duration of the radio surgical procedure:	21 hours
Duration of the anhepatic time	5-hours, 30-minutes

[a]According to the TNM Classification of Tumors.
[b]Assessed by the Galactose Elimination Capacity (GEC) Test.
[c]Normal values > 70%.

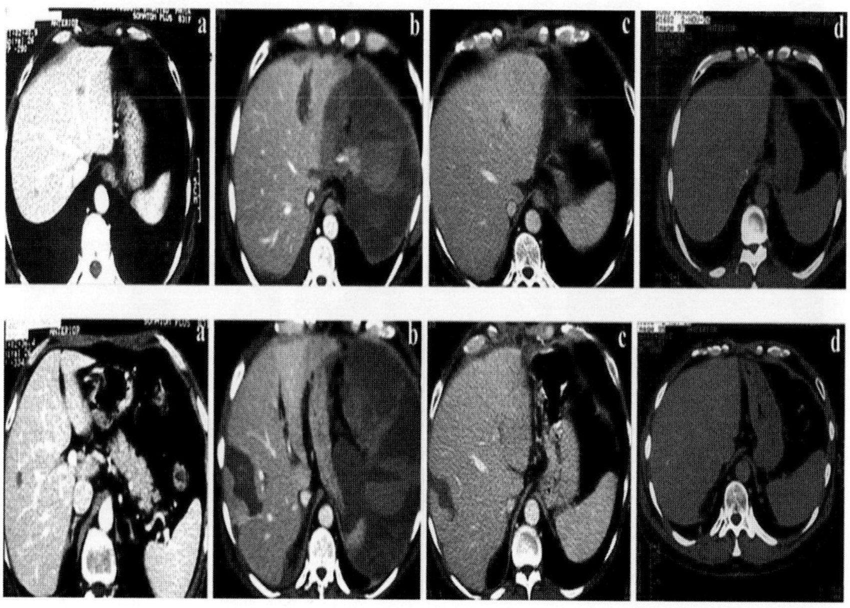

Figure 8.13 Sequence of CT images of the liver on a cranial (*above*) and a caudal (*below*) level in the first patient subjected to BNCT. Evolution at different times of the metastases towards necrosis with final substitution by normal hepatic tissue. (A): Preoperatively; (B): after seven days, (C): after six months; (D): 12-months after BNCT.[367]

Table 8.4   BNCT Pavia project. From the clinical record of the second patient.[367]

Sex and age:	Male, 39-year-old
Primary tumor:	Rectal adenocarcinoma p T3 G2 N1 M1[a]
Vital organ state:	Dilatative cardiomyopathy
Hepatic involvement:	11 small and large synchronous metastases
Residual liver function[b]:	58%[c]
Vascular abnormalities	Right hepatic artery from superior mesenteric artery
Dose of 10 BPA	300 mg/kg body weight through a colic vein
Duration of the radio surgical procedure:	18-hours, 14-minutes
Duration of the anhepatic time	6-hours, 10-minutes

[a]According to the TNM Classification of Tumors.
[b]Assessed by the Galactose Elimination Capacity (GEC) Test.
[c]Normal values > 70%.

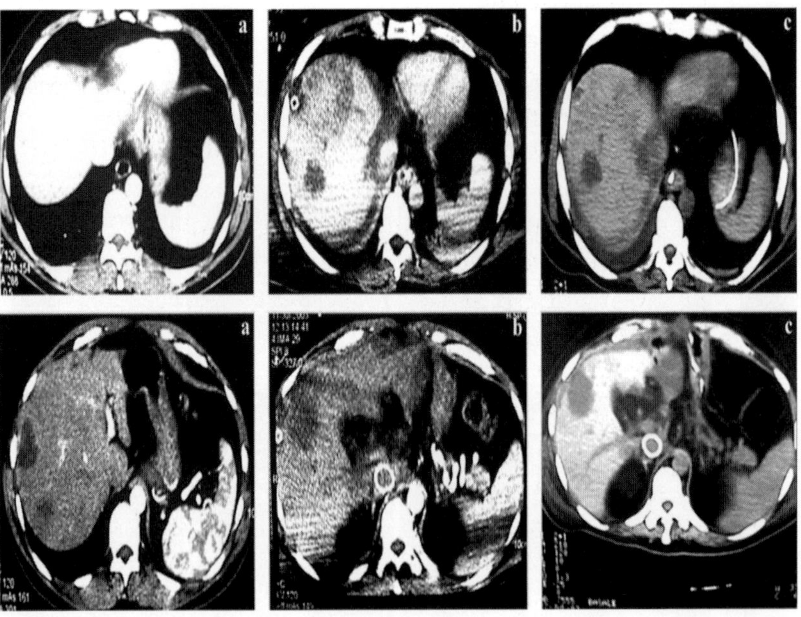

Figure 8.14   Modification of CT images of the liver on a cranial (*above*) and a caudal (*below*) level in the second patient from (a) the preoperative aspect up to the situation, (b) after 10 days, and (c) after 21 days during the peri-operative follow-up after BNCT. In the (b) and (c) images of the upper row, an area of necrosis is encircled, not corresponding in the preoperative scan (a) to any known metastasis.[367]

BNCT has also been applied in liver cancer treatments where liver autotranplantation is not possible. Suzuki *et al.* reported a case study in which a 60-year-old man with multiple hepatocellular carcinomas (HCCs) was enrolled as the first patient in a pilot study for treating multiple liver tumors with BNCT.[369] Although the BNCT-treated tumors showed regrowth 3.5 months after BNCT and the patient died of liver dysfunction caused by progression of HCC 10 months after BNCT, the feasibility of BNCT for HCC was indicated in this first case.[369]

There has been interest in the use of BNCT for tumors in radiosensitive organs, such as liver and lungs. Since the radiation damage in BNCT depends on the ^{10}B concentration, a steep concentration gradient of ^{10}B between tumor and healthy tissue would limit radiation to the tumor, sparing normal tissues. This is quite different from photon therapy, where radiation damage is distributed equally between tumor and healthy tissue. We have already discussed the use of BNCT in the treatment of liver cancers; in addition, there has been some initial reports on the use of BNCT in the treatment of diffuse or multiple pleural tumors.[370] Two patients, one with malignant pleural mesothelioma (MPM) and another with a malignant short spindle cell tumor, were treated by BNCT at KURRI. The prognosis of MPM has been dismal, and the mean survival time is only 9–12 months without intervention. The first patient (with MPM of the left chest wall) was given two BNCT treatments. In the first, the patient was infused with 250 mg/kg BPA-*f* and irradiated with neutrons 15 min after finishing the BPA infusion. In this treatment, the upper part of the tumor was irradiated with anterior and posterior epithermal neutron beams at the KUR reactor. The doses delivered to the tumor volume ranged from 10 to 30 Gy-Eq; the maximum dose delivered to the left lung was 6.5 Gy-Eq. One month later a second BNCT, three port irradiations, with anterior, posterior, and left epithermal neutron beams, was performed to treat the lower portion of the tumor, following the administration of BPA at a dose of 500 mg/kg. The dose delivered to the tumor volume ranged from 10 to 60 Gy; the left lung volumes receiving ≥ 7 Gy-Eq was 29% and ≥ 10 Gy-Eq was <1.0%. Chest pain disappeared the day after BNCT; follow-up CT at 1 and 6 months after the second BNCT showed partial regression of the tumor. Consolidation in the left lung appeared 1 month after the second irradiation and lasted for 6 months. Radiographic findings suggested radiation pneumonitis in the region where a dose > 4 Gy-Eq was delivered; there was no sign of radiation pneumonitis in the upper portion of the left lung treated in the first BNCT. The patient died of local extension of MPM 12 months after the first BNCT.

The second patient was a 43-year-old man with a 7 cm diameter lung tumor in the left lung. He received surgical resection of the tumor and histological examination revealed a malignant short spindle cell tumor. The first BNCT was performed at the JRR-4 research reactor following a dose of 500 mg/kg BPA-*f*. Three tumors in the left lower lung was defined as the gross tumor volume and treated with BNCT. The ^{10}B concentration in the tumors during irradiation was 33 ppm. The mean and maximum doses of radiation delivered to the tumors ranged from 16.2 to 32.7 Gy-Eq and from 25.1 to 44.7 Gy-Eq, respectively. The maximum dose delivered to the left lung was 11.4 Gy-Eq. Several months later three tumors located in the upper portion of the left were treated with a second round of BNCT. The ^{10}B concentration in the tumor was estimated to be 51 ppm, the mean and maximum doses in the tumor ranged from 5.8 to 13.3 Gy-Eq and from 8.2 to 32.7 Gy-Eq, respectively. The maxium dose delivered to the left lung was 5.6 Gy-Eq. Back pain, which was a major complaint disappeared within a few days after the first BNCT and follow-up CT 1 and 3 months after the first BNCT revealed regression of the tumors. The tumors in the upper portion, treated in the second BNCT remained stable in size at 3 months after the second BNCT. All tumors were enlarged within 7 months of BNCT. He died of the extension of the tumors 18 months after the first BNCT. The main concern of the two studies was to assess the extent of lung toxicity. However the results showed that BNCT was an effective treatment to palliate the symptoms of MPM. At this time neither BNCT, nor any other treatment, is curative for MPM.

8.2. Clinical Trials Using Gadolinium Neutron Capture Therapy (GdNCT)

At present, there are no clinical trials involving Gadolinium Neutron Capture Therapy (GdNCT). There are a number of gadolinium complexes approved as MRI contrast enhancement agents. In general, these complexes can be macrocyclic or linear, as shown in Fig. 8.15. Even though some of

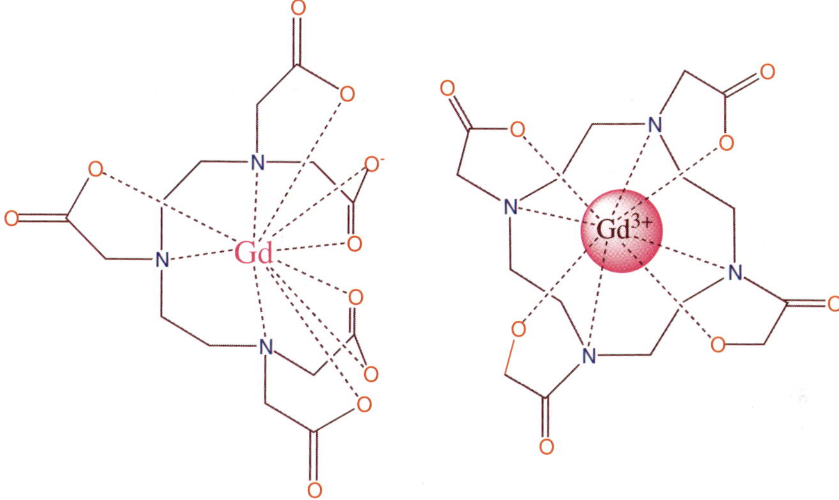

Figure 8.15   Examples of macrocyclic (*left*) and linear (*right*) Gd-complexing agents.

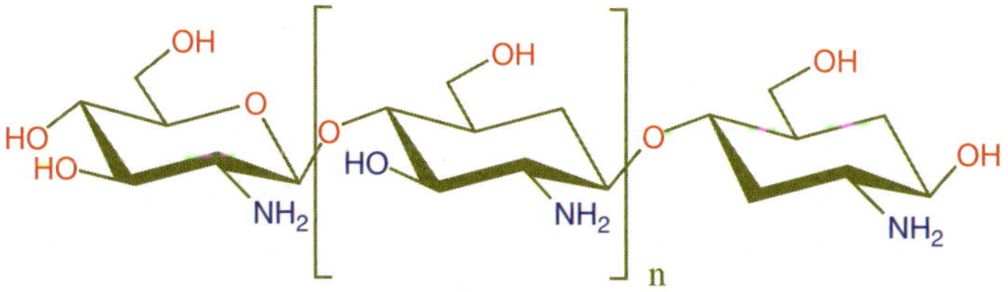

Figure 8.16   Structure of chitosan.

these can penetrate cell walls, they are not useful, since they tend to wash out quickly. Their residence time in tumors can be extended by attaching the gadolinium complex to nanoparticles, such as chitson (see Fig. 8.16).

In vivo studies, in which Gd(DTPA)-loaded chitosan nanoparticles (see Figs. 8.15 and 8.16) were intratumorally injected twice into mice bearing subcutaneous B16F10 melanomas and then irradiated with thermal neutrons 8 hr after the second injection, showed significant suppression of tumor growth compared to the group receiving only Gd(DTPA).[371] In addition to treating cancers, NCT has been applied to other medical conditions.

Restenosis is a major problem after balloon angioplasty and stent implantation. To prevent restenosis after percutaneous transluminal coronary angioplasty (PTCA), a variety of drugs have been investigated[372]; as yet, none has proven effective. In animal studies, gene therapy has been successful in preventing intimal hyperplasia or cell proliferation, but the clinical effectiveness of this treatment remains unknown. Interesting results have been reported using beta emitters to inhibit restenosis after PTCA.[373]

Although much of the interest in NCT has focused on the study of the effectiveness of NCT for treating malignant brain tumors, we believe that this technology also has tremendous potential for inhibiting the cell proliferation that leads to restenosis. A study designed to test the effectiveness of GdNCT in inhibiting intimal hyperplasia or thrombosis of vascular walls, and whether or not it causes collateral tissue damage, was undertaken. At their origins, the internal carotid and coronary arteries are susceptible to atheromatous plaque formation, which is liable to lead to clinically significant stenosis and possible lethal cerebral or cardiac thrombosis (see Fig. 8.17). While such constrictions can be

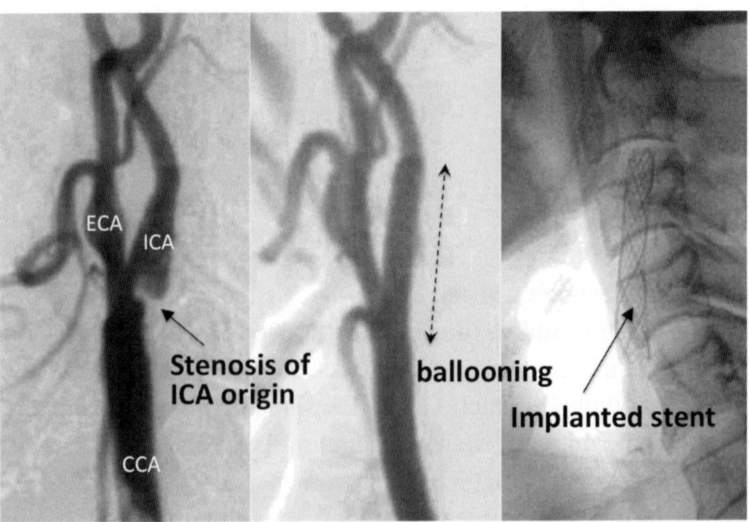

Figure 8.17 *Left.* Carotid bifurcation with severe stenosis at the origin of the internal carotid artery (*arrow*). *Middle.* Balloon catheter inserted in stenotic artery to enlarge the constriction. *Right.* After enlargement, a stent is implanted to prevent early restenosis.

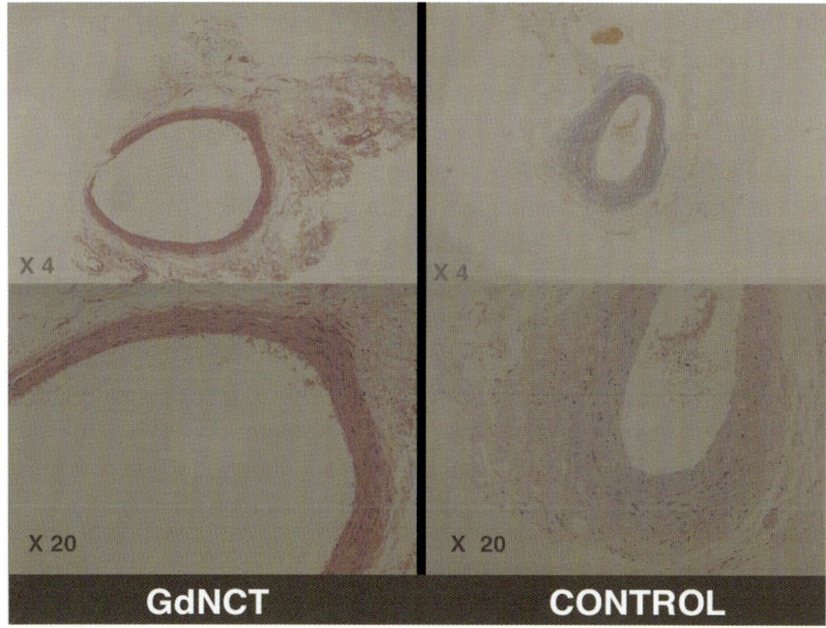

Figure 8.18   Pathological findings for the arterial walls of the control group (*right*) and the GdNCT group (*left*) of a rat model of injury-induced femoral arterial stenosis.

opened, for example, by insertion of a stent into the stenotic carotid artery (carotid artery stent: CAS), there is a high risk of restenosis at the ends of the implanted stents. Various studies are being undertaken to prevent restenosis after CAS.[374] Recent results of tests using GdNCT, shown in Fig. 8.18, indicate that it might be an effective therapy to prevent vascular stenosis.[375] As can be seen in Fig. 8.18, neither intimal hyperplasia nor thrombus formation was apparent in the GdNCT group on rat models of femoral arterial stenosis, slight intimal hyperplasia of the injured arteries was observed in all the control samples. In the GdNCT group samples, elastic lamina was almost completely preserved. This finding is important because priteoglycan sulfate, which protects against thrombosis, is present in this internal elastic layer. No venous thrombosis was observed in either group.[375]

Gadolinium neutron capture brachytherapy (GdNCB) is a suitable treatment modality for stenosis. In this treatment, a Gd-containing stent is implanted in the stenotic artery by intravascular maneuvers and subsequent radiation with thermal neutrons initiates a form of intravascular brachytherapy (IVBT).[376] This leads to selective and higher doses to the surrounding tissue and may prevent stent edge and in-stent restenosis.

GdNCT using Gd needles has been proposed, in this modality Gd-containing needles are implanted in tumors and then irradiated *in vivo* with neutrons.[377] Monte Carlo calculations show that 5000 cGy of prompt gamma dose could be delivered to a treatment volume of 40 cm^3 with a three-plane implant of 9-Gd needles. Dose measurements of a Gd needle in air also show that Gd is promising for this form of brachytherapy.[377]

Both BNCT and GdNCT are versatile forms of directed radiotherapy, which have many advantages over standard photon therapies. However, before NCT can be used in a medical setting, a number of improvements must be made. Some of these are discussed in the next chapter.

Chapter 9

Future Perspectives for Boron and Gadolinium Neutron Capture Therapies in Cancer Treatment

Neutron capture therapy (NCT) is potentially a surgically precise method of treating cancers and other medical conditions. When Locher first introduced the idea in 1936,[1] its elegance and basic simplicity captivated the minds of physicists, chemists, and physicians. Unfortunately, much of NCT still remains an ideal rather than a reality. The idea is both simple and demanding. It involves preferentially localizing compounds containing either boron (^{10}B 19.9% na, σ_{Th} = 3837 b) or gadolinium (^{157}Gd 15.6% na, σ_{Th} = 255,000 b) in tumor cells and then irradiating the particular area with low energy, thermal neutrons. For effective BNCT, the required concentration of ^{10}B is generally estimated to be 10^9 ^{10}B atoms per cell, which translates to approximately 35 μg ^{10}B/g of tissue.[11b] Under these conditions, about 85% of the radiation damage arises from the neutron capture reaction; the surrounding tissue should contain no more than 5 μg of ^{10}B/g of tissue. For GdNCT, the situation is more complex, and due to the extremely large neutron cross section, there is a significant shielding of deep tumors. To overcome this, the optimum ^{157}Gd concentration in tumors should be around 200 ppm ^{157}Gd (<1000 ppm ^{n}Gd). Since the radiation in GdNCT is tiered, the dose for effective GdNCT is very location dependent; the highest effect is found when the Gd penetrates the cell nucleus. The Gd should also have a long enough retention in tumors for effective GdNCT. De Stasio, *et al.* studied the microdistributions of the MRI contrast agents, Gd-DOTA, and Gd-DTPA in human cultured GBM cells. They intracerebrally implanted C6 glioma cells in rats (Gd-DOTA), and tumor samples from GBM patients injected with Gd-DTPA.[333] It was found that, in cell cultures, Gd-DOTA and Gd-DTPA were found in 56% and 84% of the cell nuclei, respectively, in rat tumors Gd penetrated the nuclei in 85% and 47% of the tumor cells. However, in human GBM tumors, only 6.1% of the cell nuclei contained Gd-DTPA. The efficacy of Gd-DTPA and Gd-DOTA as GdNCT agents for GBM was predicted to be low.

There have been a number of summaries of BNCT, by Barth,[379–381] and others,[382] and a comprehensive review on both BNCT and GdNCT by the present authors.[11a] In discussions associated with what is needed to advance BNCT and GdNCT, the following points have been considered to be severe stumbling blocks in the efforts to accomplish that goal.

(1) Currently, the US Food and Drug Administration Agency (FDA)-allowed boron drugs (BPA, BSH, and GB-10) and gadolinium drugs (Gd-DOTA, Gd-DTPA) are neither tumor-specific nor homogeneously accumulated in tumor cells.

(2) For clinical trials, the patients have to be transported to one of a limited number of nuclear reactor facilities that are currently available internationally, thus restricting the ready availability of the treatment for the needy patients.

(3) Restrictions imposed by the FDA on more clinical trials involving new boron and gadolinium compounds, including the novel Gd-doped boronated magnetic nanoparticles, have limited the clinical efficacy of the BNCT/GdNCT in cancer treatments.

(4) Restricted research funding for the BNCT/GdNCT studies by the US National Institute of Health (NIH) and other health-related agency has had a negative impact on the numbers of researchers in the area in the academic, clinical, and medical professions.

(5) Without proper media publicity, it has been an impossible task to bring about public awareness on the existence of the state-of-the-art cancer treatment option through BNCT and GdNCT.

Foremost is the need for bringing improved boron delivery agents into clinical trials. The boron chemistry has reached a stage, demonstrated by the awards of two Nobel Laureates in Chemistry (William N. Lipscomb, Jr in 1976 and Herbert C. Brown in 1979), where the question is what to build rather than how to build. The minimum ^{10}B concentration for effective BNCT has been estimated to be 10^9 ^{10}B atoms per cell, for a mono-boronated agent, such as BPA. This translates into roughly a 1–3 mM concentration, compared to the nanomolar and micromolar concentrations of other chemotherapeutic agents. Given that BNCT is based on the differential absorption of B in cancer cells, high boron concentrations must be avoided in healthy cells. This imposes rather stringent requirements on a boron delivery agent, which must concentrate heavily in tumor cells while largely avoiding healthy cells. Hawthorne and Lee[382] suggested a four-group classification of boron delivery agents: (1) global boron agents with no cell targeting; (2) boron agents having malignant cell recognition only; (3) agents that bind only to cell nuclei; and (4) those agents having tumor cell recognition and binding to the nucleus of the targeted cell. At present only three boron compounds are in clinical trials, GB-10 ($Na_2B_{10}H_{10}$), BSH, and BPA. BSH and GB-10 belong in category one, while BPA is a category-two drug. These drugs are the result of chemical research carried out in from 1958 to 1965. Among all the new boron delivery agents developed in the last 50 years, none have made it to clinical trials. As pointed out earlier, the results obtained on these two boron compounds are not universally promising; they present low tumor-to-blood and tumor-to-brain tissue ^{10}B ratios.[82–88] Unless other, more efficient, boron carriers are brought to chemical trials, it is difficult to project appreciable advancement in this field.

The second need is to move away from nuclear reactors as the sources of neutrons. Such sources are not conducive to a hospital setting. Unless accelerator-based, or small-fusion-based, neutron generators are developed, BNCT will always be an expensive, rarely applied treatment. The technology for construction of such alternate generators is known. Tanaka *et al.* found that a cyclotron-accelerated 30 Mev proton beam of 1 mA impinging on a Be target would generate a neutron beam having superior characteristics to that of present day reactors.[284] However, unless the US Department of Energy (DOE) develops more efficient and portable neutron delivery accelerators, applications of BNCT and/or GdNCT in cancer treatment facilities within the United States will not be practical. Nonetheless, the current situation may change rapidly if some spectacular tumor-specific boron or gadolinium compounds were to become available for clinical trials. Until then, the availability of portable neutron sources will be low or nonexistent.

The exclusive use of nuclear reactors is also considered to be a bottleneck in the development of new compounds for NCT. New delivery agents can be developed in chemistry laboratories and *in vitro* testing could be done in a laboratory setting. In some cases, biodistributions in small animals can be measured. However, it is very difficult to carry out the next logical step, that is, to test tumor inhibition under neutron irradiation. Nuclear reactors are not that accessible. What is needed is a system of grading the results of *in vitro* and *in vivo* studies and moving forward promising compounds to neutron sources for further testing. This would involve the setting up of clearance centers for assessing the results of *in vitro* and *in vivo* testing and for planning further studies of promising compounds. This will no doubt require government sponsorship of nuclear testing facilities. As it now stands, only a favored few who have access to reactors can move their candidate drugs forward.

A greater emphasis should be placed on treating cancers other than brain cancer (GBM). While it is true that GBM is essentially untreatable using standard methods, it is also true that, despite years of testing with BSH and BPA, BNCT is not a cure at present. At best, BNCT might extend the time to regrowth, but it has not yet been found to be curative. The process of BNCT is complex, but has not been demonstrated to be superior to standard treatments for GBM. Linz in a recent publication has questioned the wisdom of pursuing BNCT as a cure for GBM.[384] With the presently available boron delivery agents, the question is a valid one. GBM is one of the most difficult cancers to treat, it is protected by the BBB, and is extremely infiltrative. On the other hand, there have been very encouraging results reported for other cancers. It is an open question whether progress in the cure of a few other cancers, for which alternate therapies might be available, would be enough to warrant the large financial investment necessary for routine BNCT.

The prime question for the medical profession to ask would be whether there is any future for boron and gadolinium neutron capture therapy (BNCT and GdNCT)? The conclusions can be drawn as follows:

(1) For as long as there is no significant tumor control, let alone cure, for a number of fatal cancers is available through current therapies, there will be a need for new therapy approaches, such as NCT. NCT in the forms of BNCT and possibly GdNCT are based on a solid binary treatment concept, which has every chance to be successful with the intense drug design and optimization research that is currently underway. More specific and selective NCT agents will invariably lead to improved boron or gadolinium carrier dosimetry and better treatment planning software. The more that is known about the complex medical and pharmacological requirements for NCT at the atomic, molecular, microbiological, physical, and biophysical level, the better the treatment standards will become. *Tumor specificity and selectivity, along with the homogeneity within the tumor cells of the drug, remains a central issue not only for NCT but also for the majority of other treatment modalities.*

(2) It would make sense to combine NCT or neutron capture-enhanced fast neutron therapy (NCEFNT) with proton therapy, for example, within future Hadron Therapy Facilities.[385] It remains to be seen whether the combination of the neutron capture reaction and fast neutron tumor bombardment produces the anticipated radiation dose enhancement. *Again, drug specificity and its homogeneity in the tumor cells remains a key issue.* The use of combined boron and gadolinium carriers, such as Gd-doped boronated magnetic nanoparticles, is certainly attractive, especially with regard to dynamic drug uptake monitoring with MRI. A combination of the two NCT reactions is also feasible with the use of NCT drug cocktails, combining individual BNCT and GdNCT agents. Currently, this approach also appears useful from a pharmacological point of view as a way of seeking a boosted therapeutic effect exploiting the BNC and the GdNC reactions within a single boron-gadolinium carrier.

Thus, the development of more effective delivery agents, such as boronated and Gd-doped magnetic nanocomposites,[386] is the key requirement in the advancement of BNCT and GdNCT. If these developments are encouraged through federal funding, with more clinical trials and public support, the BNCT/GdNCT research will attract new talented researchers into the field, and, eventually, alternative neutron sources will be developed. If not, BNCT and GdNCT will remain as the intriguing but elusive therapy that is just over the horizon.

Appendix

The Clinical State of Boron Neutron Capture Therapy was compiled in 1997 by the United States Department of Energy (DOE) — Office of Science under Office of Biological and Environmental Research & Medical Sciences Division and can be downloaded free of charge (http://www.er.doe.gov/ober/73_reports/bnct.pdf) for personal use without prior permission from DOE or the Publisher of this book. This material is reproduced by kind permission of the Office of the Assistant General Counsel for Technology Transfer and Intellectual Property at the United States Department of Energy, granted June 9, 2010.

United States Department of Energy
Office of Science
Office of Biological and Environmental Research
Medical Sciences Division

Report on the Workshop:

The Clinical State of

Boron Neutron Capture Therapy

1997

Park Hotel, Charlotte, NC, November 3–5, 1997

Workshop Organized and Report Compiled
by
Ludwig E. Feinendegen

Published December 1998

Medical Sciences Division
Office of Biological and Environmental Research
Office of Science
U.S. Department of Energy
19901 Germantown Road
Germantown, Maryland 20874–1290

Telephone: (301) 903-3213

Set in Galliard (Carter & Cone)

Title page set in Miller (Carter & Cone; Font Bureau)

Table of Contents

Summary

The Department of Energy (DOE) held a workshop on 'The Clinical State of Boron Neutron Capture Therapy (BNCT)' in Charlotte, North Carolina, November 3–5, 1997. The workshop was aimed at assessing the present state of ongoing clinical trials, to guide future planning, and to prepare for decisions. A preceding DOE workshop on BNCT held May 9-12, 1995 in Williamsburg, Virginia, addressed 'Research Needs for Neutron Capture Therapy.'

BNCT uses boron-10 labeled compounds that accumulate preferentially in a selected target such as tumor cells. The target is then exposed *in vivo* to thermal neutrons which induce the $^{10}B(n;\alpha)^7Li$ reaction. This instantaneous nuclear reaction deposits about 2.5 meV locally over a range of about 15 μm, i.e., 1-2 cell diameters, and is lethal to the affected cells. Thus BNCT is considered as a unique technique for accomplishing non-invasive 'targeted and timed cell surgery' in the living body.

Thirty-three participants represented the main US research groups engaged in or immediately preparing for BNCT clinical research. Institutions represented included:

- Brookhaven National Laboratory (BNL) with the Memorial Sloan Kettering Cancer Center,
- Various campuses of the University of California with the Lawrence Berkeley Laboratory,
- Idaho National Engineering & Environmental Laboratory (INEEL),
- Massachusetts Institute of Technology with the New England Deaconess-Beth Israel Medical Center of Harvard University (MIT/H) ,
- Ohio State University,
- University of Tennessee Medical Center,
- Washington State University.

One participant from Neutron Therapies Incorporated (California) is presently general secretary of the International Society of Neutron Capture Therapy. Four participants were from the Department of Energy.

The agenda focused on expectations and challenges of BNCT based on the data from the ongoing phase I clinical trials at the Harvard-MIT program (15 patients) and phase I/II clinical trials at Brookhaven National Laboratory (35 patients). Also, the Japanese BNCT experience was reviewed. The presentations and discussion included treatment planning and dosimetry with special focus on normal tissue tolerance and also included consideration of BNCT as an adjuvant to fast neutron therapy. Further discussed were potential new compounds for clinical BNCT with reference to *in-vivo* assessment of compound biodistribution and to the optimal delivery and targeting of compounds to tumor cells. Potential application of BNCT to malignancies other than glioblastoma multiforme and melanoma and to nonmalignant diseases were identified. Finally optimal criteria for selection and follow-up of patients were addressed.

The summarizing general discussion acknowledged the safety and lack of unusual adverse effects of BNCT at dose levels used with the present protocols for malignant brain tumors (45 patients) and for cutaneous malignant melanoma (5 patients). The continuation of the ongoing trials, therefore, considers escalation in the amount of boron-10 labeled compound (boron-phenylalanine, BPA) and also escalation of tissue dose from the epithermal neutron flux, as approved by the Institutional Review Boards and the Food and Drug Administration. The maximum absorbed dose of 12.5 Gy-equivalent to the normal total brain should not be exceeded. The ratio of doses to tumor and normal brain is expected to exceed 4. Also, the potential of BNCT as adjuvant to fast neutron therapy was considered promising.

The mean survival times of the patients with glioblastoma multiforme treated at Brookhaven were 15 months in protocol group 1, and more than 10 months to date in groups 2 and 3; the corresponding times for conventional therapy are 10.5 months in group 1 and similar to those seen so far in the other groups. More patients need to be evaluated for statistical data analysis. In 5 patients with BNCT of cutaneous malignant melanoma, again no toxic or adverse effects were seen. The treated cancer nodules in the skin responded well, and, depending on size, were seen to fully disappear with no recurrence so far in one patient 2.5 years after BNCT. The normal skin showed lower radiation effects than are normally seen after conventional radiotherapy. The convenience of a single BNCT treatment, versus multiple treatments over about two months required by current conventional radiotherapy, was considered to improve the quality of life.

Based on these data, further clinical trials for the purpose of treatment optimization are justified. These trials are expected to involve optimization of the BPA dose and evaluation of retargeting techniques, i.e., extending from a single treatment session to repeated sessions—probably two, one day apart—so that targeting of boron-compound to tumor cells may be improved. No compound other than BPA is foreseen for clinical use in the near future in the USA. Yet, biodistribution studies with BSH may be justified as preliminary to eventual dual compound application with BPA for BNCT. Moreover, interesting new boron-10 labeled compounds as metabolic analogs and with new delivery systems are being developed for preclinical testing and perhaps for imaging purposes in humans. Neutron dosimetry needs adjustment for multiple beam projections in order to spare normal tissue.

While the primary focus at present is on BNCT protocols for brain tumors and malignant melanoma, plans are being developed for clinical application of BNCT of affected joints in patients suffering from rheumatoid arthritis within the next 2–3 years at the Massachusetts Institute of Technology. Basic research is advancing well. Other malignant and non-malignant diseases were identified as potential later candidates for BNCT according to disease type, location, and responsiveness to other therapy modalities.

The following report also summarizes statements from the session leaders whose engagement and contributions are gratefully acknowledged.

BNCT Expectations and Challenges

After more than 3 decades of basic research, early clinical trials of BNCT resumed in the United States in 1994 for malignant melanoma and glioblastoma multiforme at the Tufts University Medical Center and Massachusetts Institute of Technology (MIT) in Boston, and for glioblastoma multiforme at the Brookhaven National Laboratory (BNL). These trials are intended to primarily demonstrate safety of the proposed therapies, and are designated as phase I clinical trials by the Food and Drug Administration. Although these trials have the potential to reveal some measure of efficacy, the database is not sufficient as yet to do so. Two such clinical trials with dose escalation have been completed at BNL with a total of 35 patients. They show that BNCT appears safe during the follow up period at the utilized levels of neutron dose and boron-phenylalanine (BPA). Also, the patients whose glioblastoma multiformes were treated with BNCT lived as long as the patients treated conventionally. Finally, BNCT treated patients appreciated the option of a single treatment vs. the multiple treatments required by conventional therapies. These initial results confirm the expectations raised two and a half years ago at the DOE workshop held in Williamsburg. While these results are encouraging, no statement regarding clinical efficacy can now be made.

The trials in the United States are expected to continue. The phase I clinical trial will be completed in one to two years. Phase II trials with BPA will continue during the next two to three years, and are likely to allow assessment of effectiveness at some level of tumor dose. Moreover, a phase I trial involving glioblastoma multiforme has recently begun in Europe, at the Petten reactor in Holland; this trial uses sulfhydryl duodecaborane (BSH) with retargeting in 4 fractions, one day apart. In Japan, the use of BNCT for skin melanoma and brain tumor will continue as discussed later in this report.

If a clear indication of efficacy evolves from phase II trials (compared to conventional control treatments), phase III trials are expected to begin in order to definitely ascertain the effectiveness of BNCT in peripheral melanoma, in brain metastases of melanoma and in glioblastoma multiforme, or in a subset of these malignancies. Other clinical phase I and phase II trials of BNCT may use advanced boron-10 labeled compounds.

It is hoped that the next few years will bring an answer to the applicability of BNCT in clinical practice not only with regard to treating glioblastoma multiforme and melanoma but other malignant and non-malignant diseases as well. A number of questions were raised with regard to the conduct of future clinical trials:

- What lessons have been learned from the on-going BNCT trials?
- Should a trial of BNCT be permitted as a retreatment after a full course of chemotherapy and/or radiotherapy, and if so when? The present recommendation precludes BNCT as retreatment.
- To what degree and when does tumor surgery alter the blood brain barrier and local cerebral perfusion? This may affect the biodistribution of i.v. injected target compound.
- What target compound will likely succeed the presently used BPA and BSH? The differential uptake ratio of tumor/normal tissue appears sub-optimal for both compounds. Desirable new compounds should concentrate in tumors at more than 100 ppm giving a tumor to normal tissue concentration ratio of more than 10. Such new compounds should be retained in the tumor for several hours and be non-toxic.
- How can one assure homogeneous distribution of sufficient quantities of target compound in the tumor? The eventual success of BNCT depends on essentially 100% tumor cell kill. Does retargeting of tumor cells through fractionated injections of compound optimize the trial protocol?
- What is the optimal fractionation schedule of neutron irradiation in case of retargeting of tumor cells with compound? The spectrum of radiation qualities in the target body at the time of neutron irradiation of the tumor causes different acute or late effects.
- What is the best treatment plan for assuring effective depth dose distribution in the brain? Tumor recurrence was observed in regions with relatively low local dose. Does the dosimetry need to assure irradiation of larger volumes of the brain than the tumor volume, so that infiltrating tumors are properly exposed?
- Which software is optimal for effective therapy planning? Different planning procedures and software are in use in the BNL and Harvard-MIT programs; they need to be compared for data evaluation. Agreement on therapy planning will advance the on-going trials.
- What is the optimal unit of radiation dose for realistic use in BNCT? The presently used units are not uniformly accepted and do not comply with ICRU standards.
- Are BNCT protocols now ready for intercomparisons and inter-institutional data analysis? The continuation of BNCT trials will benefit from pooling experiences and data.
- Is a common referral base for BNCT patients in the USA or internationally now desirable? Other therapy trials operate with such referral bases.

- What other malignancies and non-malignant diseases are likely candidates for BNCT?
- Are ethical questions relating to BNCT trials fully explored and understood? Compassion for patients and realistic assessment of potential benefits of BNCT must be carefully balanced.

Many of the answers to the above question are essential contributions coming from this workshop. Most of the participants agreed with the need to complete phase I clinical trials with BPA in the USA within one or two years. Similarly, participants endorsed the desirability of phase II trials for BPA during the next 2–3 years. There was also no disagreement that BNCT effectiveness at some level would be demonstrated prior to the completion of the clinical trials. However, it remains to be seen if an adequate indication of effectiveness will emerge to justify the initiation of a phase III trial. A doubling of the minimum survival time for glioblastoma multiforme was suggested as decisive evidence of BNCT effectiveness and an adequate justification to proceed to a randomized phase III trial. Pre-clinical trials using BNCT are foreseen for non-malignant diseases.

The proper clinical follow-up of patients by their primary care physicians in close cooperation with the medical group that provides a trial was seen indispensable for long term and reliable evaluation of data. The costs of these follow-ups should be adequately incorporated into the accounting of the overall costs of the clinical research programs. Autopsies are particularly valuable and everyone involved should make all efforts to obtain at least partial autopsies when treated patients die of whatever cause.

For the purpose of optimally evaluating compounds that carry elements for capturing thermal neutrons, the concept of a 'clearing house' was discussed. This should also serve to coordinate the introduction of new compounds for neutron-capture-therapy. Presently, the evaluation even of one class of compounds is very demanding and the limited resources make this a daunting task.

The following research and development is foreseeable for the next few years:

- Microscopic imaging with high resolution track etch autoradiography and surface physics techniques combined with Monte Carlo evaluation of data will provide needed information on boron microdistribution and kinetics. This will lead to optimal treatment planning with neutron irradiation.
- Treatment planning systems in use by most groups will be sorted, cross calibrated and compared. This will allow an intercomparison between the different clinical programs. The time required for completion of treatment planning may be reduced to less than one hour and dose delivery may be possible with a spatial resolution of about 5 mm.
- A few promising new compounds labeled with boron-10 or another useful

element for neutron capture therapy will enter extensive preclinical testing.

- A few reactor based improved epithermal neutron sources and one or two accelerator sources of epithermal neutrons will be ready for preclinical and clinical trials.

Presentation of Clinical Data

a) The Harvard/MIT experience:

The presentation of data of the clinical BNCT trials presently conducted at the Massachusetts Institute of Technology and the Beth Israel Deaconess Medical Center of Harvard University (MIT/H) began with an expression of appreciation to the Brookhaven National Laboratory (BNL) staff, which provided MIT/H with BPA-fructose (BPA-F) on several occasions and instructed MIT/H staff members to synthesize BPA-F. The review from MIT/H featured first the five patients who were treated for peripheral malignant melanoma on the skin of a leg or foot. The total target doses given in 4 daily fractions in four patients and in one patient in 1 exposure ranged from 10 to 12.5 RBE-Gy The single field neutron irradiation was started at 30–60 minutes after beginning of BPA-F infusion, 400 mg/kg body weight for the each of the four irradiation fractions and 250 mg/kg for the single irradiation. The boron-10 concentration ratio for tumor to blood ranged from about 2.5 to 3, with the skin having boron-10 concentrations similar to that of the peripheral blood in the various patients. Two patients showed partial response at 11 and 17 months after BNCT, two had complete response including the patient with a single irradiation schedule who was followed up to 34 months (as of the Workshop); one patient could not be completely followed. The complete remissions in two patients were confirmed by histological examination of biopsy tissue that was taken from the treated sites. The normal tissue was initially red with dry desquamation soon after BNCT, but at the time of biopsy was completely normal and elastic.

These first results indicated that BNCT of malignant melanoma of skin was safe. Also, no significant chronic normal tissue reactions to radiation occurred in any of the patients followed. All patients showed partial tumor response with two having a complete remission up to nearly three years after BNCT, despite applied radiation doses which were well below the threshold for significant normal tissue reaction.

Because of many competing protocols for developing treatment for brain tumors in the Harvard hospitals, the number of patients referred there for BNCT of glioblastoma multiforme is still relatively small. So far, nine patients with glioblastoma and one with melanoma metastasis to the brain have undergone BNCT at MIT/H. The study protocol escalates radiation dose to normal tissue and brain from 8.8 to 12.9 RBE-Gy. At the time of the workshop the 10.6 RBE-Gy level had been completed. The minimum tumor dose was about 2 RBE-Gy. Delayed radiation effects to normal brain tissue did not show upon repeat examinations with magnetic resonance imaging (MRI) or upon clinical examination. The patient with metastasis of melanoma in the occipital brain region had complete tumor regression 9 months after BNCT. In the other patients, tumor recurred after 3 to 13 months following BNCT, with a median survival time of close to 9 months. It should be noted that therapy after BNCT was not standardized.

For therapy planning and dosimetry, pharmacokinetic data were acquired by sampling venous blood during the time of irradiation. The samples were analyzed for boron-10 by conventional analytical techniques such as the prompt-gamma procedure. In consenting patients, stereotactic biopsies of tumor and adjacent brain tissues were obtained following a test dose of BPA-F. Intracellular boron-10 concentrations were measured by high-resolution quantitative autoradiography. Generally, 25 ppm average of boron-10 were measured in the peripheral blood at the end of the one hour infusion into the jugular vein with a rapid decline thereafter, whereas brain tumor tissues at the time of neutron irradiation had boron-10 concentration about 2–3 times that found in blood. On the basis of such data and the corresponding MRI's, dose-volume histograms were constructed for each of the 10 patients treated on the brain tumor BNCT protocol.

The ensuing discussion emphasized that about 30 RBE-Gy is considered the minimum necessary radiation dose to the brain tumor for effective BNCT. Yet, with as much as 110 RBE-Gy to the tumor about 75 % of the patients still had tumor recurrences. This may be due to the fact that the present BNCT protocol plans for four daily fractions. Depending on the dose reduction factor for fractionated exposure in BNCT of glioblastoma multiforme, a total dose of 110 RBE-Gy given in four daily fractions could be equivalent to a single dose of about 22 RBE-Gy. Another crucial factor in this evaluation is the boron distribution to tumor cells at the time of neutron irradiation. This was again referred to later in this workshop. For optimal efficacy, all tumor cells must have a boron uptake sufficient for cell killing upon the n-capture.

One 58 years old patient with a posterior-parietal brain tumor suffered a thalamic infarct after the first treatment was given. This infarct developed into a fatal event. No radiation induced injury to normal brain was evident on MRI, but massive edema around the irradiated, partially resected large tumor developed and is believed to be the cause of the patient's demise. This patient was admitted to BNCT with what was described a marginal condition, although each inclusion criterion had been met.

The minimum, average, and maximum radiation doses to tumor were estimated for each patient for different exposure modes. As an example, the resulting treatment plans for the patient with a melanoma metastasis to the brain compared the tumor dose obtained from the M-67 beam at the MIT-reactor with that from the fission converter beam (FCB) being under construction there. The latter is expected to increase the deliverable dose rate by a factor of two. As a consequence, with normal brain receiving the tolerance dose of 12.5 RBE-Gy, the minimum dose to tumor from the FCB will be about 40 RBE-Gy, which is well above the desired minimum of 30 RBE-Gy for inducing a remission of glioblastoma multiforme.

b) *The Brookhaven experience:*

At the time of the workshop, 35 patients had received BNCT at the Brookhaven Medical Research Reactor (BMRR); 34 of these patients can be evaluated. They are divided into four groups as shown in the table below.

Group	No. of Subjects	BPA-F Dose mg/kg	Reactor Power (MW)	Radiation Fields	Collimator Diameter (cm)	Median BNCT Dose (Gy-Eq) Average Brain	Minimum Tumor
1	14	250	2	1	8	2.2	27
2	8	250	3	1	12	3.6	38
3	5	290	3	1 or 2	12	5.0	44
4	7	250	3	2	12	4.8	28

Groups 1, 2, and 3:

One of the 14 patients in group 1 was alive without any evidence of tumor recurrence at 21.5 months after diagnosis was made, and 20.6 months after BNCT. Five of eight patients in group 2 were alive and in one of these five patients the removal of the necrotic tissue from the tumor bed at 11 months after BNCT did not show any evidence of residual tumor. All five patients in group 3, the most recent group with the shortest follow-up, were alive and in good clinical status.

The median values for age, Karnovsky Performance Score (KPS), Curran Class, tumor volume and tumor depth in the 14 patients in group 1 were 57 years, 85, 5, 21 cm³ and 5 cm respectively. The median survival time was 15.1 months after the diagnosis. The patients in groups 2 and 3 were clinically similar to patients in group 1; their median survival times could not yet be determined. The median post-BNCT follow-up times for groups 2 and 3 at the time of the Workshop was 11.3 months and 4.5 months respectively.

Group 4:

The seven subjects in group 4 belonged to a two-field exposure study of BNCT. The aim was to determine the feasibility of offering BNCT to patients with larger and/or deeper tumors as compared to tumors in groups 1,2, and 3; also at test was reasonable safety with the potential for delivering a tumor control dose. The results suggest the two-field exposure may be as safe as the single-field exposure. However, the potential of the epithermal beam at the BMRR to deliver a tumor control dose to larger and/or deeper tumors was limited by the relatively inadequate flux of thermal neutrons at depth in brain. Further trials for tumors that are located deep in the brain and/or are large are being postponed until the proposed fission-plate convertor is installed at the BMRR.

The median age in this group was 48 years, the median KPS was 80, the respective Curran Class was 4, and the median tumor volume and tumor depth were 37 cm^3 and 6.8 cm respectively. The median survival of group 4 patients from diagnosis to the date of the Workshop was 11.5 months, and two of seven patients were alive at 11.5 and 11.8 months after the diagnosis.

Clinical and pathological findings:

Acute side-effects were observed in almost all patients following BNCT. They included alopecia, erythema, transient lymphocytopenia and transient granulocytosis. Moreover, when the irradiation field included an ear, temporomandibular joint (TMJ), parotid gland, conjunctiva or buccal mucosa, the observed grade 1 or grade 2 temporary side-effects included otitis externa, otitis media, parotitis, conjunctivitis, tenderness of TMJ and change in taste. All these side-effects were successfully treated by conventional medical management.

Seizures occurred in 7 patients after BNCT and were promptly controlled in all patients by conventional anti-seizure therapy. Only one of these 7 patients suffered the transient seizures as a direct consequence of BNCT.

Two patients with vasogenic edema around the residual tumor prior to BNCT also developed increased intracranial pressure at about 12 hours following BNCT. In both instances, an increase in the dose of dexamethasone was therapeutically effective.

No late side-effects were recorded after BNCT in any patient up to the time of the Workshop, but since tumor recurred in all at various time intervals, with associated clinical deterioration, late side effects of BNCT were hard to judge. Autopsies were done in five patients who had been followed from 4.5 to 14.6 months after BNCT. Histopathological examinations of the brains of these five patients failed to show any BNCT induced damage to normal brain. Tumors recurred locally in one half of the BNCT patients, and locally plus regionally in the other half.

The Brookhaven data may be summarized as follows:

- The BPA-F infusions did not cause any acute toxic effects.
- Acute BNCT side-effects generally were transient and responsive to standard medical interventions
- Autopsy evaluation of normal brain in five patients found no histological damage to the normal brain after BNCT.
- The life expectancy of BNCT treated patients has been extended a few months beyond what would be expected following surgical debulking alone and is in the range of that seen in patients treated with conventional radio- and/or chemo-therapy after surgery.
- Monitoring patients enrolled in groups 2 and 3 may yield some dose-effect relationships regarding tumor regression.
- The gross tumor BPA-F uptake correlated with tumor cellularity using morphometric techniques. This correlation justifies the use of a 3.5:1 tumor/blood boron concentration ratio for estimating tumor doses.
- Calculations indicate the possibility of treating large and deep brain tumors with BPA-F and the proposed improved beam (fission plate convertor).

The results of these safety-driven protocols led to the proposal to move on to optimization of future protocols. This will help to determine the potential of BNCT in eventually providing local tumor control. The proposed studies plan to deliver radiation doses close to the limits of tolerance of the normal brain; amendments are planned for the case of fractionated irradiation, likely to entail 2 fractions. In order to account for any sensitization effects induced by the first fraction, the total dose requires adjustment to a value about 10% below the brain tolerance dose. The protocol to be prepared in the very near future will include the following studies:

- Dose escalation to normal brain tolerance employing single-fraction irradiation with one or more fields is planned for such tumors which are located laterally in cerebral hemispheres but not in anterior-temporal and anterior-inferior frontal lobes.
- Two fractions of irradiation with two or more radiation fields will be chosen for tumors located in anterior-temporal and in anterior-inferior frontal lobes. Soft tissues such as the eye, ears, and salivary glands do presently not permit the delivery of an adequate dose in a single-fraction to tumors located in the given locations. This study will employ graded dose escalation, 3 patients per dose level.

- An exploratory study plans the inclusion of glioblastoma multiforme tumors which have not been debulked, have a volume of less than 20 cm³, are located in accessible regions and are otherwise eligible for BNCT. The tumors must be proven by biopsy, and neurosurgeons shall clarify that gross debulking poses an unacceptable risk. These non-debulkable tumors will be treated with two fractions. The first dose will be less than that in the main dose escalation protocol for 2 fractions. At each dose level, three patients will be studied before proceeding to the next level with about 15% dose escalation at each step.

c)   The Japanese experience

After the early BNCT trials in the USA were discontinued in the 1960s, BNCT as a form of treatment for malignant gliomas moved to Japan under the direction of Dr. Hatanaka in Tokyo. This program is now under the direction of Dr. Nakagawa, following the death of Dr. Hatanaka. This treatment program uses sulfhydryl-duodecaborane (BSH). A second program in Kobe applies BNCT with boron-phenylalanine (BPA) mainly for malignant melanoma of the skin, under the direction of Dr. Mishima.

The BNCT of malignant melanoma in Kobe is well documented in the literature and has resulted in some tumor regression sand total remissions.

Early reports on the results of BNCT of brain tumors showed surprisingly high survival rates. This helped stimulate resumption of clinical work in the USA. However, only recently has there been an attempt to critically analyze the Japanese experience. This is important in the light of cultural differences that lead to different standards for clinical reporting.

Consequently, the data for the 14 patients from the USA who received brain tumor BNCT in Japan entered a review conducted by Drs. A. Sperce (neurologist/neuro-pathologist) and Laramore (radiation oncologist). The analysis remained limited to this subset of patients because it was felt necessary to both review medical records for important prognostic factors and to obtain tumor pathological specimens for central review. One patient turned out to have a brain lymphoma and, thus, was excluded from the analysis. Of the 13 patients in the analysis, two had anaplastic astrocytomas and 10 had glioblastoma multiforme. Patients were classified according to the recursive-partitioning analysis of Curran, et al. On a pseudo matched pair analysis, the BNCT treated group had no therapeutic benefit. Median survival for the glioblastoma patients was 12 months. All have died prior to the time of the Workshop, and all had tumor recurrence and some also brain necrosis with concomitant cerebral failure. The only long-term survivors were the two patients with anaplastic astrocytomas, per-formance classes I and II, who would have had a high survival probability also with conventional forms of treatment.

This work must be put in perspective relative to the BNCT trials in the USA. The treatments in Japan used thermal beams and BSH given intra-arterially; the tumor to blood concentration ratio remained between 0.5 and 1.0. Moreover, the thermal beam required an open craniotomy, and led to an irradiation time of 6–8 hours. The current programs in the USA circumvent most of the difficulties inherent in the Japanese setup. Currently, work in Japan has shifted to the use of epithermal treatment beams from two reactors and emphasizes the treatment of high grade gliomas and malignant melanomas.

Treatment Planning and Dosimetry

a) Primary and secondary radiation units in BNCT

* *Base and derived quantities and units in* BNCT:

The presently applied dosimetric units for BNCT are not easily transferable to the units common in conventional radiotherapy; they also do not comply with the recommendations of the International Commission on Radiation Units and Measurements (ICRU). Thus, the concept of absorbed dose was discussed first.

The quantity 'absorbed dose,' at a given point, is the relevant quantity in radiation therapy and commonly used for the different modalities of irradiation such as in external beam therapy with photons, electrons, neutrons, protons, etc. The absorbed dose, at a point, is defined as the 'mean energy imparted' per unit mass, in a volume surrounding that point. The averaging process implies that a large number of ionizing particles cross the defined volume of interest (ICRU Report 33). A strict relation exists between absorbed dose and cell lethality, and it varies with different types of radiation.

The situation is more complex in BNCT where different types of radiation contribute to the absorbed dose at a given point; thus, the following dose components arise:

* gamma rays from the reactor and from the hydrogen capturing neutrons;
* the 'nitrogen dose,' which is the absorbed dose from the capture of thermal neutron by nitrogen-14 atoms giving rise to accelerated protons and carbon-14 atoms;
* fast neutrons;
* the so called 'boron dose,' which is the absorbed dose from the capture of thermal neutrons by boron-10 atoms giving rise to accelerated alpha particles and recoiling lithium-7 atoms ranging together over about 13–15 μm, i.e., a cell diameter.

The concept of absorbed dose can be applied to the first three components as is done in the conventional therapeutic radiation modalities, since the energy depositions occurs stochastically within the neutron radiation field. However, the concept of absorbed dose has limitations in the case of the boron dose, since the incorporation of boron-10, thus the number of alpha particles and recoiling lithium-7 atoms, is not uniform in the tissues in the neutron radiation field. In fact, the boron concentration varies from one cancer cell to another one depending on the type of boron-10 labeled substrate, the tumor type, cell metabolism and local blood supply. Under such conditions, the 'averaging process' implied in the definition of the concept of absorbed dose is no longer meaningful.

- *Modifying factors:*

In clinical situations, it is only possible to average the boron concentration over a large population of cells and thus derive a kind of average dose. If tumor cells do not contain boron or contain less boron than normal tissue, any increase in neutron fluence will yield no therapeutic benefit but will disproportionately increase the toxicity in normal tissues. Therefore, so-called Compound Factors or modifying factors may only allow a therapeutic gain calculation, if the boron-10 distribution in the tumor regarding cells and extracellular space is known. In case of less boron in a given tumor cell than in the extracellular space, a negative therapeutic gain would result with respect to normal tissues even if the average boron concentration in the tumor over normal tissues may indicate a significant gain.

In normal tissues, the same difficulties apply to some degree. However, the boron concentration in blood of a given patient can be determined accurately and repeatedly during irradiation and the distribution of boron may follow a relatively predictable pattern. Known RBEs and Compound Factors from experiments may be used to establish a starting dose for phase I dose escalation studies. The ultimate tolerance has to be determined by careful dose escalation studies where particular attention must be paid to the possibility of unexpected boron uptake in a sensitive normal cell population in the exposed tissue.

- *Recommendations*:

According to the foregoing discussion, at least three aspects need to be considered in BNCT:
  - fractionation of the irradiation (number of fractions, dose per fraction, overall time);
  - number of administrations of boron-10 labeled compound;

- timing of drug administration(s) and irradiation(s).

BNCT consisting of a single fraction neutron irradiation following a single administration of boron-10 labeled compound can not be considered as optimal, even if the time interval has been optimized (ratio of blood vs tumor boron concentration). The experience accumulated in conventional radiation therapy over many decades has proven the benefit of fractionation: no tumor has been cured, or can be cured, by one or a few radiation fractions. Even in fast neutron therapy, e.g., high LET irradiation, the TAMVEC experience has shown that 2 fractions per week could be dangerous and that at least 3 fractions per week are needed. Also in chemotherapy, a large number of repeated drug administrations are required, and no example testifies to the efficacy of a single drug administration. Although the situation of BNCT as a binary form of therapy is different involving, for example, both active and passive transport of the boron compound in the tissue, it seems to be much safer to follow the general rules currently and successfully applied in radiation therapy, in chemotherapy and in combined radio- and chemotherapy.

The 'non-boron dose' components, see above dose components, significantly contribute to the dose to the cancer cells, as well as to the normal tissues and thus should be optimized.

Regarding dose escalation, three aspects need to be understood:

- escalation in exposure time;
- escalation in the amount of boron administered;
- escalation in both factors.

Escalation in exposure time will proportionally increase the contribution of the four radiation components (see above) contributing to the absorbed dose at the point(s) of interest. Also, the normal tissues at risk will receive higher doses and they will come closer to their tolerance limit.

As far as the boron dose in concerned, a therapeutic gain arises only when the boron concentration is principally higher in all the cancer cells than in the normal tissues at risk. In contrast, if the boron concentration in some cancer cells is lower than in the normal tissue cells at risk, increasing the exposure time will result in a negative therapeutic gain. This is true, of course, in any case with some cancer cells having not incorporated any boron at all.

The same arguments likely apply to any single administration of larger amounts of boron compound because of given alterations in local tumor blood supply, cell metabolism, effects of previous treatments. In contrast, an improved therapeutic gain could be expected with increasing the number of boron administrations properly spaced in time. This would result in a more homogeneous boron distribution in the tumor than is expected from one administration alone, since repeated administrations increase the probability of the boron compound to reach all the tumor cells.

Regarding the question of retreatment, the discussion led to generally advise against it according to the extent that the normal tissue tolerance in BNCT is approached.

Uniform definitions of terms and concepts, as well as the same dosimetric protocols should be used by all clinical centers involved in BNCT. This is essential for relevant exchange of information, data analysis, and for evaluating the outcomes of the various trial phases.

In addition, it was stated to be important that the BNCT community adopts, each time it is possible, the general terms and concepts currently used by the radiation therapy community in general. This should crucially help to exchange information, to compare the results and simply improve credibility. In that respect, the ICRU definitions of volumes, and the ICRU recommendations for reporting the treatment, with regard to the specification points, should be followed and adapted and supplemented when necessary, to the specific situation of BNCT.

b)   Computational dosimetry and treatment planning for BNCT

- *Status of the relevant technology:*

For approximately ten years the DOE has supported the development, maintenance, and deployment of two independent software systems for computational dosimetry and treatment planning of BNCT. One of these systems has been developed by the Idaho National Engineering & Environmental Laboratory (INEEL) with clinical collaboration from the Brookhaven National Laboratory (BNL). The other was initially developed at Tufts/New England Medical Center and is currently maintained in connection with the MIT/H clinical BNCT program. Both of these systems are successfully used to support human clinical BNCT trials under FDA-approved clinical protocols. This is a remarkable achievement considering the fact that BNCT needed a new expertise for performing the complex, mixed-field dosimetry calculation, with patient geometry constructed directly from the relevant medical images.

It is to be emphasized that even though the two software systems are presently used at both BNCT clinical trial sites in the USA, the work of the two different development groups is by no means redundant. For instance, the INEEL/BNL system has been designed from the beginning with much broader applications, beyond the use of an epithermal neutron beam. Indeed, the DOE may expect to be in the position of having supported the development of a computational tool that will prove useful in all fields of neutron radiotherapy as well as possibly for some non-neutron applications.

Even if the two separate DOE-sponsored dosimetry software systems were totally redundant, the dual system strategy of the DOE for developing these tools was successful and should be continued. The cost is relatively modest in the overall BNCT effort; moreover, several significant innovations emerged, both as a result of occasional 'friendly competition' between the two development groups as well as from synergistic cross-fertilization of good ideas. For example, the INEEL/BNL system of radiation dose computation has often featured significantly faster execution speeds than the MIT/H system. This is a result of the the former system concentrating on such algorithms that were specifically written for medical neutron transport applications, while the the latter system, for quite valid reasons, preferred a standard, but slower, general-purpose Monte Carlo Program (MCNP) for the dose computations. More recently, however, the MIT/H group created a special version of the MCNP with an improved execution speed by adopting the INEEL/BNL experience with certain improvements and advancements in geometric representation and particle-tracking algorithms. Conversely, because of the decision to build the INEEL/BNL geometric reconstruction algorithm to be independent of image-modality, the MIT/H system now has a greater degree of computer-automated reconstruction of patient geometry when used with CT. Developers of the INEEL system have had the opportunity to learn from the MIT/H team.

In summary, in terms of the requirements and expectations established at the Workshop, the tools for computational dosimetry and treatment planning of BNCT are becoming well-established. One system or the other, and in many cases both, now can meet most of the current expectations; significant improvements are expected in execution speed, clinical user-friendliness, and breadth of application, e.g., regarding fast-neutron therapy with adjuvant BNCT. The participants considered BNCT to be in a successful stage of development with definite advancements in therapy planning.

- *Issues related to specific applications of computational dosimetry in BNCT*

The participants supported the cross-correction of results between the INEEL/BNL and the MIT/H computation systems. An excellent opportunity to do this comes from the developing research at the MIT on BNCT for synovectomy that will probably lead the INEEL/BNL system to be licensed to MIT in 1999. Also, the MIT/H group now validly normalizes the dosimetry calculations to in-phantom measurements. The INEEL/BNL group, on the other hand, successfully prefers a stage by stage validation of the complex computational sequences involved; the objective is to achieve *a-priori* consistency of theory and measurement in the final results. Both approaches are clinically valid.

The participants also addressed the issue as how to define the tumor and target volumes when performing dosimetry calculations. Much of this is a matter of personal preference and experience. Within a fairly broad range of what is considered by the radiation oncologists to be acceptable, different treating physicians will likely maintain different views on how to define these regions for prescribing an optimal therapy.

c) *Single dose tolerance of normal brain*

The tolerance of the normal brain tissues exposed to BNCT is obviously the main limiting factor to the tumor doses. The BNL group has utilized a rat spinal cord model and the thermal beam at the BMRR to define the relative biological effectiveness (RBE) of the primary and secondary beam components and the neutron capture fission reaction. The model uses paralysis as an endpoint. The Seattle group used the epithermal beam at the BMRR for testing the response of the normal dog brain as to the RBEs for the various beam components; the two endpoints were brain lesions visible by magnetic resonance imaging (MRI) in otherwise healthy animals, and severe neurological dysfunction from brain necrosis appearing about 5 months after treatment and leading, in fact, to a rapid death. On MRI, tissue lesions appeared after 9 Gy, whereas neurological symptoms became obvious after about 12 Gy. The results obtained have been used to set the doses for additional studies regarding both multifraction exposure and retreatment.

The discussions of these data attempted to define tolerance in the normal human brain. The human brain is much larger, and the treatment field is larger. The use of a 1 cm³ volume brain for setting the maximum dose was considered overly conservative since the dog tolerance was derived from the response of about 20% of the brain volume, i.e., at least 30 cm³. The results observed in normal brain tissue of the animals are now used in the treatment planning for human patients.

The estimated RBEs and the so-called Compound Factors expressing the biological effectiveness of the boron-compound used, are relatively uncertain; but the sum of all the factors appear to be reasonable.

The following factors appear justified:

- RBE
  Fast neutrons = 3.3
  Nitrogen capture protons = 3.3
  Incident and capture gamma radiation = 1.0
- Compound Factor
  BPA = 1.3
  BSH = 0.33

d) *Multifraction irradiation studies*

In their work with the epithermal beam at the BMRR, the Seattle group again used the dog brain model and split the total dose into 2 and 4 equal fractions given at 24 hour intervals. Tissue repair following the low LET components did not appear in these studies and the results were identical to single dose response. There was an increased skin reaction following the fourth fraction compared to single dose exposure. The BNL group using the rat spinal cord model also compared the single fraction to 2 and 4 fractions given at 48 hour intervals. While a slight amount of repair could here be measured, this was much less than observed in a gamma irradiated control group. The presently inevitably low dose rate in BNCT may negate the repair that is now included in computational fractionation.

The discussion resulted in a general consensus that the human trials should start fractionation by dividing the total single dose into equal fractions and not assume any repair initially. Also, because the soft tissue response may be worse after fractioned irradiation, the proposed fractionation schedule should not involve all fields equally, but the irradiated fields should change to optimally spare normal tissues. The main purpose of the fractionation scheme in BNCT remains the optimal compound distribution within the tumor in terms of retargeting to potentially reach all tumor cells.

e)   Reirradiation tolerance

The Seattle group presented results from seven dogs that received BNCT twice. First treatment involved the administration of 250 mg BPA/kg followed by epithermal neutron irradiation at the BMRR. The estimated dose to about 20% of the brain volume, i.e., to about 30 cm^3, was 11.5 Gy in three dogs and 12.5 Gy in four dogs. These doses were expected to result in no lethal outcome and close to 50% MRI changes based on previously published single-fraction work. After six months, the dogs were treated again with the same dose to the same region. All seven dogs developed lethal lesions. The time interval between the second irradiation and the onset of pathological symptoms was unexpectedly short; in three dogs neurological findings appeared at approximately two months. It was concluded from these studies that the damage from the first irradiation was not repaired with the consequence of a shorter than usual time interval until onset of lethal lesions after the second irradiation.

In rat experiments, the BNL group gave three different doses of 6 meV photons to a 2 cm long segment of cervical spinal cord. After six months, BPA was administered intravenously and the same region of the spinal cord received various BNCT doses to determine the dose effective at 50 % level, the ED_{50}, at the time of retreatment. For the groups that had initially received 22, 40, or 80 % of tissue tolerance, the retreatment ED_{50} were 77, 80, and 50% of tolerance, respectively. This set of data from rat studies, thus again, showed a sparing effect with indication of repair.

An additional group of rats received initially BNCT and showed at six months a remaining 55% of tolerance. These BNCT retreatment results were graphed with other published data on rat spinal cord using photons or mixtures of fast neutrons and photons. The BNCT data and the published data from fast neutron/photon experiments fit the generalization that the retreatment tolerance depends on the magnitude of the initial dose. If the initial dose was about 50% of tolerance, the retreatment tolerance of the normal brain is reduced to about 80%. If the initial treatment dose was 80% of tolerance, at retreatment the remaining tolerance is only about 50% of the ED_{50}.

The ensuing discussion emphasized the uncertainties on how to relate the animal data to the human situation. For example, the rat spinal cord model provides no information on volume effects. The volume of the dog brain is about 150 cm^3, thus much larger volumes received higher doses than one schedules in patient treatments. It was also pointed out that the tolerance of the normal human brain to BNCT is unclear. How to fold volume effects into the tolerance data available from published data on clinical photon irradiation, experimental BNCT, and the current clinical BNCT trials is an area needing further investigation. The discussion also addressed the magnitude of the tumor dose that could be delivered in BNCT as retreatment. Since normal brain tolerance obviously limits the dose for retreatment, and if this consideration reduces the total tumor dose to below the value of a single dose, serious ethical questions arise. The general opinion of the participants tended to assemble more preclinical and single fraction clinical BNCT data before the recommending BNCT as a retreatment option after failed photon therapy.

f)   Clinically implemented dosimetry

A review gave the design and operation of the INEEL treatment planning system for clinical studies at BNL. It emphasized how the shapes of the computed isodose curves depend not only on the neutron beam spectrum but also on the specific dose components and their mix present at a given position in the target tissue. Thus to be considered are the dose components discussed above, namely from the boron neutron capture process, the gamma-dose from beam contamination and neutron capture reaction predominantly with hydrogen ($^1H(n;\gamma)^2H$, the dose from the neutron capture reaction with nitrogen ($^{14}N(n;p)^{14}C$), and the dose from fast neutrons reacting mainly with hydrogen nuclei. The calculations considered the following factors of boron concentration in relation to the circulating blood: in brain 1, in the scalp 1.5, in mucosae 2.0, in the tumor 3.5. The respective Compound Factors, as discussed above, were taken to be 1.3 for the brain, 2.5 for the scalp and mucosae, and 3.8 for the tumor. The RBE for other high LET dose components attained the value of

3.2. Dose-volume histograms were shown to be good computational tools for examining the biological impact of various treatment plans. The INEEL treatment planning program could be interfaced to a 3-D anatomy and isodose display module developed at BNL.

In summary, the following general issues need consideration in programming of treatment planning:

- 3-D dose distribution in all body parts, with sufficient resolution, perhaps of 5 mm, and with separate dose components regarding the RBEs to be validated by experiments,
- accuracy of dose estimates optimally 5 % but not less than 20 %, using proper transport media relating to different body structures, shapes and sizes,
- modeling based on x-ray computed tomography (CT), MRI, and functional imaging with nuclear medicine techniques, preferentially in a fused mode,
- normalization of data to those obtained by use of phantom experiments,
- arbitrary plane isodose display that is superimposed on the corresponding anatomy,
- dose-volume-histograms, giving volumes, and the maximum, minimum and average doses,
- interface to calculate the prescription of exposure regarding the MW power of the reactor and time of irradiation.
- adequately fast computational facilities, in a stochastic, deterministic or hybrid mode to be decided.

For the processing of treatment planning, the following general issues appear summarily essential:

- informed consent of patient, and all legally required forms and signatures,
- treatment room availability for pre-therapy acclimitization of patients to the particular setting,
- availability of all necessary images for prescribing the individual treatment, with written directives containing tumor and target region contouring, beam placement, computed dose, evaluated plan, beam choice and vector to patient, and on-line assays of boron-10 concentration in peripheral blood,
- availability of baseline clinical tests,
- assurance of sterility and absence of pyrogen in the boron compound solution to be injected.

The absorbed doses delivered at BNCT at BNL were established by using the treatment planning software package developed at INEEL. High spatial resolution of this system allows the creation of detailed models of body support devices, as well as patient's heads including anatomical structures, e.g., tumor and target volumes, as well as various regions of the brain such as cerebral hemispheres, cerebellum, brain stem, basal ganglia, and optic chiasm. These models are then used in Monte Carlo (MC) calculations, which combine the patient geometry with neutron beam characteristics and body elemental cross-sections, locally absorbed energy distribution, RBE and Compound Factors to produce three-dimensional dose distribution. Results of the MC calculations can then be displayed as isodose contours of total dose or any major BNCT dose component over an arbitrary plane of the MRI. Also, dose-volume histograms for structures defined in the model and any dose component can be drawn, and minimum, maximum and average doses can be obtained. It takes approximately 1.5 hrs of computer time to calculate one case by the current version of the software. This will be upgraded early next year reducing computer time to less than 30 min. The results obtained from the INEEL system were verified by measurements as well as independent calculations using the MCNP-MC code. Continuous close collaboration between BNL and INEEL leads to further improvement of the system and adjustment to clinical needs.

Accordingly, the doses deliverable in BNCT at BNL may be summarized as follows:

- The dose to the contralateral hemisphere can be minimized through careful treatment planning even when two-field irradiation is applied. A third field could be applied from the contralateral side to further increase the dose to deep parts of the tumor,
- the gradients of various dose components are different and their relative contribution to the total dose varies with depth. This may have radiobiological consequences to be incorporated in the treatment plans for the fractionated mode of irradiation,
- the dose-volume histograms, average doses and their components for various brain structures are different in two-field irradiation; for example, more than 50% of the dose delivered to the contralateral hemisphere derives from gamma radiation. This dose distribution can be exploited for the fractionated mode of irradiation,
- increasing the normal brain peak dose in dose escalation studies seems to be safe because only a small fraction of the normal brain will receive doses higher than those that were proved to be clinically safe. Moreover, the fraction of the normal brain, which will receive the highest dose is located within the target volume, a region which comprises the tumor and a 2 cm tissue shell around it. These high-dose volumes will be comparable to the volumes receiving doses as high as 15 Gy in patients treated by conventional stereotactic radiosurgery.

The analysis of doses to the normal brain tissues in BNCT patients treated so far supports the proposed dose escalation and fractionation for the next protocols.

g) BNCT: *an adjuvant to fast neutron radiotherapy*

Fast neutron radiotherapy alone or as adjuvant to treatments has shown its efficacy in more than 30,000 patients with cancer. Numerous, randomized clinical trials have been conducted and neutron radiotherapy gave better local control than conventional radiotherapy for advanced salivary gland tumors, prostate cancer, and sarcomas. For other tumors such as squamous cell tumors of the head and neck and non-small cell lung cancer, the results are equivocal, and for tumors such as glioblastoma multiforme there is no evidence of therapeutic efficacy. Because of the limited tolerance of tissues, the tumor doses can not be significantly increased beyond their current values. A neutron capture therapy boost can selectively increase the tumor dose and has the potential of dramatically improving tumor control.

As fast, i.e., high energy, neutrons pass through tissue they spontaneously produces an attendant cloud of 'slow,' i.e., low energy or thermal, neutrons that are effective in neutron capture reactions. This thermalized component of the fast neutron radiotherapy beam has been analyzed at the University of Washington Medical Center (UWMC). Model calculations predict a ten to one hundred fold increase of local effectiveness in killing of cells with a boron-10 concentration in the range of 30 μg/g. This prediction has been verified with *in vitro* measurements on the U79 cell line, and *in vivo* in the 36B10 rat glioma model; the concept showed its validity also in a human melanoma patient using BPA as the boron-10-carrier.

Non-small cell lung cancer has been selected for the next phase of study. A prior randomized trial showed an apparent therapeutic benefit for the subset of patients with squamous cell lung carcinoma; also, the analysis of the pattern of failure indicated the spinal cord to be the dose limiting organ in many cases. According to plan at UWMC, the borane $^{10}B_{10}H_{10}$ will be used which is not a metabolite nor tumor specific but which is excluded from the cerebral spinal fluid. However, an appreciably high blood/tumor ratio needs consideration in determining the allowable augmented dose to the spinal cord. Animal studies confirm an extremely low pharmacological toxicity and indicate a potentially high tumor boron concentration (~100 ppm).

Present work at the UWMC tests the efficacy of a BNCT boost in the fast neutron therapy of spontaneous lung tumors in dogs. The next steps aim at validating the toxicity and kinetics of the borane concentration in the blood in normal human volunteers. Then, tumor uptake of the compound will be investigated in patients with non-small cell lung cancers and glioblastoma multiforme, at the time of surgical resection. Institutional Review Board (IRB) approval has been obtained and forwarded to the Food and Drug Administration (FDA) for final endorsement to the holder of license of an Investigational New Drug (IND). Eventually, Phase I/II trials will use functional and structural changes in PET images for assessing treatment efficacy, as a surrogate endpoint for survival.

Compounds for Clinical BNCT

The design of the overall BNCT program must include continuing studies in chemistry for preparing optimal compounds for biological, preclinical, and clinical work such as that now in progress in the USA and abroad. Specific concerns address the design of boron compounds and their *in vivo* delivery to target cells; the latter issue is often overlooked. If research funding were abolished or even decreased in these areas, improvement of BNCT would be restricted to neutron sources and biology, manipulation and modifications of clinical use of the two compounds of the 'first generation', boron-phenylalanine (BPA) and sulfhydryl-duodecaborane (BSH). Furthermore, there is no physical or medical reason that BNCT must be limited to glioblastoma multiforme and melanoma nor confined to modifications of the present protocols with epithermal neutrons. The $^{10}B(n,\alpha)^7$ Li reaction is a unique binary source of localized high LET particles and its potential applications in medicine are limited by available resources rather than ideas.

New therapeutic applications arise from the synergistic matching of the neutron source with the best available boron compound for a given disease under attack. This demands the definition, emphasis and support of certain critical areas of research. Three areas of investigation relevant to the chemistry of compounds for BNCT are: compound design, synthesis and evaluation.

a)   *Compound design*

Compound design is based upon the differential performance of tumor cells and normal cells including the vasculature which supplies these cells. The type and metabolic characteristics of the tumor to be treated and the selected neutron source can be dominant parameters.

In order to achieve an optimal targeting and retargeting, the synthesis and biochemical/biological evaluation presently focusses on low molecular weight boron compounds. Whatever the compounds, they should adequately concentrate boron-10 in ideally all cells of the target tissue to a value of about 10^9 boron-10 atoms per cell, or approximately 30-35 μg boron-10 per g tissue average. At the same time, the ratio of boron-10 concentrations in target tissue and normal tissue such as in the tumor to normal brain, and the corresponding ratio of concentrations in tumor and blood must ensure high radiation doses to the tumor while the maximal tolerated doses to normal brain and the vascular endothelium are not exceeded. In addition to targeting the main tumor mass that has not been excised, it is essential that the compounds have the capacity for crossing the normal blood-brain barrier (BBB). This is essential for targeting those tumor foci that have infiltrated normal brain. These tumor islands must also be destroyed since they could be the basis for tumor recurrences and thus account for therapeutic failures.

At present, BPA and BSH are two old and fairly well understood agents in clinical trial; they belong to the 'first generation' of compounds and are not optimal for several reasons. One is the lack of tumor selectivity of both compounds, with BSH being concentrated in brain tumors for their lack of blood brain barrier, whereas BPA is actively transported across the blood brain barrier and concentrates in the tumor because of an increased rate of protein synthesis in the tumor cells; it is, however, not incorporated into protein. New and more selective compounds for given target tissues need to be developed and carefully evaluated for clinical application. Several compounds have been synthesized and are known to cross the BBB as does BPA. They need further evaluation for optimal targeting of tumor cells.

Of the 'second generation' boron compounds, porphyrin carriers currently undergo animal testing and show a promising high ratio of boron concentrations in tumor and normal brain; yet, the toxicity of porphyrin is still a problem. 'Third generation' compounds include boron labeled analogues of natural small molecular precursors of cell metabolism, lipoprotein, liposomes, various DNA binders, ligands for specific cell receptors, as well as radiation sensitizers and various pharmaceuticals.

Compounds chosen for biological evaluation need, of course, to be tested for toxicity. This must be sufficiently low for further evaluation in brain tumor-bearing animals. It is essential that those compounds demonstrating promise as tumor-targeting agents must ultimately be screened in animals with intracranial lesions that simulate clinically-observed glioblastoma multiforme. All new compounds must be compared biologically against the two agents that are now used in clinical trials of BNCT, namely BSH and BPA. Any new compound must be at least as good as BSH and BPA in terms of tumor boron concentration in order to be considered for further evaluation. Such evaluation would include more extensive toxicological studies in larger animals together with radiobiological studies.

For many treatment targets, cellular characteristics are fairly well understood. However, in order to design the optimal compound with a corresponding structure/function relationship items such as hydrophilicity *versus* hydrophobicity, stereochemical consequences of introducing large substituents, disruption of hydrogen-bonding opportunities, alteration of electrical charge and dipole moments need consideration. One often sees tailored molecules equipped with elegant specificity linked to hydrophobic carborane cages or hydrophilic polyhedral borane cages without success in biological application. This failure may derive from finely tuned cell metabolism being incompatible with compounds carrying large structural units such as the carboranes. This speaks against small compounds as boron carriers unless one has Quantitative Structure Activity Relationship (QSAR) data available and the ancillary data required to use them appropriately.

On the other hand, macromolecular compounds carrying large numbers of boron atoms can be precisely assembled and attached to ligands targeted for receptor sites that are specific for, or over-expressed by tumor cells. Certain of these macromolecular carriers can be designed to function as compounds in their own right or used as reagents for conjugation to primary structures such bioligands. Thus, small liposomes with diameters of 50–100 μm and constructed of lecithin-phosphatidylcholine appear to be exceedingly effective as carriers of hydrophilic boron compounds to be delivered *in vivo* where access to the tumor cells is open, i.e., where the blood brain barrier is inactive or destroyed. The carrier liposomes may have external targeting devices although these systems have been examined with only a small number of tumor models. Also, lipophilic carboranes may be incorporated into the liposome wall. Ideally, an i.v. injection of properly tailored liposomes may deliver a boron compound to cell receptors followed by internalization of the hydrophilic boron compound into the cytoplasm of the targeted cell followed by rapid migration and binding of the compound to the cell nucleus and its DNA. Such systems are under investigation.

b)   *Compound synthesis*

Crucial, yet often overlooked, for new compound development is the identification of a source of suitable ^{10}B-enriched precursors for compound synthesis and the development of high-yield synthesis reactions which reduce costs.

Examples of potentially wide applicability are the linking of carboranes to peptides by way of SH-or NH_2 bonding or the labeling of a variety of organic molecules especially nucleosides by using particular side chains on carboranes such as through phosphate diesters. Nucleoside linked carborane phosphate diester has been found in the cell nucleus at concentrations of 1–6×10^6 boron atoms per cell.

c)   *Interinstitutional compound evaluation*

A first step in compound evaluation is the identification of representative tumors to be targeted, including glioblastoma multiforme. The kinetics of compound uptake and distribution in the tumor and its cells needs to be appropriately demonstrated *in vivo*. Also, the toxicity of the compound should be known. Animal models will provide for testing of many needed parameters to prepare a compound for clinical use even if the study of compound distribution and toxicity in humans is essential. Little of this work is under way in a logical manner with newly developed compounds.

The basic and indispensable data on compound behavior *in vivo* should be provided through an interinstitutional project. The corresponding program would be not prohibitively expensive with defined procedures. A special group of experts should draw a program and, if a consensus is reached, additional funding should be sought for the purpose. Funding for compound evaluation without continuing support for new compound development would defeat the purpose. Obviously, the development of improved compounds for BNCT of glioblastoma multiforme or other targeted tumors should be of highest priority; trials with suboptimal compounds are in progress. It should also be recognized that boron agents are not pharmaceuticals and offer no efficacy of their own. Thus, different rules and administrative regulations apply which make compound development for BNCT less expensive and consequently easier to change due to reduced investment.

d) *Optimization of compound targeting*

An alternate way of boron compound delivery to brain tumors uses intracarotid (i.c.) injection. Thus, BSH or BPA so administered gave a double tumor boron concentration compared to values obtained following intravenous injection. This result was further improved fourfold following the disruption of the blood brain barrier (BBB) by hyperosmotic mannitol injection. In contrast, boron concentrations in normal brain and blood at 2.5 hours following i.c. injection with or without disruption of the BBB had fallen to levels equivalent to those observed after i.v. injection. These relative increases in tumor boron uptake were associated with corresponding increases in mean survival times (MST) of F98 glioma bearing rats following BNCT. The MSTs of rats given i.c. BSH or BPA were 52 and 95 days respectively when the BBB was disrupted, versus 40 and 52 days without BBB disruption. The control rats received the compounds by i.v. injection and had MSTs of only 33 and 37 days. These results demonstrate an improvement of compound delivery resulting in a significant enhancement in therapeutic efficacy in a rat brain tumor model that until recently has been incurable by any therapeutic modality. Moreover, when BSH and BPA were given together by i.c. injection into rats with a disrupted BBB, as many as 25% of the animals were cured. This strategy to optimize delivery of BSH and BPA may be applicable clinically and relatively soon be incorporated into the protocols of ongoing clinical trials. In preparing for this as a first step, boron-10 distribution in tumor cells after intra-arterial administration of BSH and BPA should be evaluated in glioblastoma multiforme patients at the time of surgical resection of the tumor. If enhanced tumor uptake of boron and the ratio of boron concentrations in tumor and blood on the one hand and in tumor and normal brain on the other are improved over the corresponding values from presently used schedules, a BNCT study should be initiated comparing i.v. versus i.c. administration of the boron compound. The disruption of the BBB by hyperosmotic mannitol is clinically used to enhance the delivery of cytotoxic chemotherapeutic agents at various institutions, including the Ohio State University. Concomitantly, studies in canines may help in assessing the safety of the procedure.

Besides hyperosmotic mannitol, the pharmaceutical RMP-7, a synthetic nanopeptide and bradykinin analogue, is used to disrupt the BBB and has been shown to enhance the delivery of BPA to a similar degree as with mannitol. Current studies with RMP-7 aim at optimizing the delivery of BPA.

Taking advantage of specific tumor cell receptors, high molecular weight ligands such as boron-10 labeled epidermal growth factor (B-EGF) will require strategies that are different from those for low molecular weight compounds such as BSH and BPA, in order to optimize boron delivery to tumor cells. Using the C6 rat glioma model, intratumoral (i.t.) injection of B-EGF resulted in a 1,000 fold increase in EGF uptake by receptor-positive C6 glioma cells compared to values obtained following i.v. injection. Intratumoral injection is being used clinically at Duke University Medical Center to deliver ^{123}I-labeled anti-EGF receptor monoclonal antibodies (MoAbs) to residual tumor cells after surgery in patients with glioblastoma multiforme. This technique or one of its variants may be applicable to deliver B-EGF or boron-10 labeled MoAbs for BNCT after primary tumor resection. Obviously, different delivery strategies appear to be needed for low and high molecular weight boron containing compounds and the development of these strategies should proceed in parallel with compound synthesis.

Besides primary targeting, potential retargeting of malignant brain tumors by boron compounds is presently an important issue in designing new protocols for optimization of BNCT. The purpose of retargeting is to assure that all cells in the target tissue take up sufficiently large amounts of boron for the capture reaction to be effective. The efficacy of retargeting depends on the degree of local tissue perfusion at the time of compound administration. Various strategies may achieve the goal of retargeting. Canine studies are planned to test for optimal fractionation of compound injections followed by repeated neutron irradiation, as discussed above.

e)   In-vivo imaging of compound biodistribution

Various modes of imaging allow the *in vivo* assessment of boron compound distribution. Particularly discussed were magnetic resonance imaging (MRI) and positron emission tomography (PET) as potent tools to quantitatively observe *in vivo* the bio-kinetics of a given boron carrying agent.

Boron MRI has not yet reached clinical efficacy suitable for BNCT treatment planning. Current protocols can generate boron-11 images. Yet, they have a relatively poor spatial resolution and, moreover, require special equipment regarding the instrument transmitter and receiver hardware; also the software needs adjustments. It is currently impossible to image boron-10 due to its poor nuclear magnetic characteristics; the atoms quadrapole leads to extremely short relaxation times (T2) making standard MRI protocols inappropriate.

Future work may allow for somewhat improved boron-11 imaging by using high field clinical research magnets, over 7 tesla; this is likely not suitable for boron-10. The latter may eventually be better imaged with conventional MRI equipment allowing the generation of new pulse sequences geared to the spin-transfer polarization between hydrogen and boron. Also, heteronuclei such as fluorine-19 may serve for compound imaging by MRI. For example, MRI using ^{19}F-BPA is a logical step; yet, the relatively large quantity of labeled compound to be administered for MRI would require toxicity testing.

PET has been used to clinically investigate uptake of fluorine-18 labeled BPA in brain tumors. So far, only three patients have undergone this examination in the USA and show distinct uptake of the labeled compound in the tumor with the tumor size appearing somewhat larger that in the MRI scans. The results confirm the reports from a greater number of patients in Japan. To date, the data look promising in that answers to important clinical questions may be obtained. Still, before treatment planning is modified on the basis of PET data, more patients must be scanned. For further development of functional imaging with PET also carbon-11 labeled BPA was prepared and the results compared to the fluorine-18 labeled BPA, at the University of Tennessee.

It is currently assumed that ^{18}F-BPA mimics BPA *in vivo* based on a limited number of animal experiments; here, nude mice were implanted with human glioblastoma tumor cells and i.v. injected with the appropriately labeled compounds for kinetic studies of tracer uptakes into the developed tumor. Although the assumption of biokinetic compatibility of ^{18}F-BPA and BPA is reasonable, further validation is needed. The presently available ^{18}F-BPA data suggest the following:

- BPA uptake can be followed from the time of injection to the *in-vivo* biodistribution, using the ^{18}F decay statistics. Interestingly, the data obtained from the three patients with glioblastoma multiforme in the USA closely resemble the data that were obtained at the BNL; here, boron concentrations were measured as a function of time during and after BPA injection, in the peripheral blood, in the tumor, and normal brain tissue, yielding uptake, and washout rates in each patient.

- ^{18}F-BPA imaging after surgery confirms that normal brain does not unusually accumulate BPA. This is also true for edema areas of the brain. However, this preliminary data was obtained after a bolus injection of the compound and does not necessarily apply when BPA is infused slowly. The images also reveal that residual tumor tissue after surgical resection takes up ^{18}F-BPA more effectively than normal tissue; the ratio of tracer concentrations in tumor and blood exceeds the corresponding ratio for normal brain and blood by a factor of more than three. The images also showed a tracer distribution that superimposed rather well with the contours of absorbed doses generated at BNL using tissue samples obtained during debulking surgery. The concurrent findings from a single brain autopsy lead to the preliminary conclusion that the post-surgery PET data may improve the BNCT planning protocol.
- PET with ^{18}F-BPA after BNCT revealed that the tracer accumulated in tissue near the original tumor boundaries. It remained uncertain whether the uptake delineated necrotic tissue or tumor regrowth. Superposition of this PET data on a gadolinium-enhanced MRI of the same patient showed the tracer to be within the gadolinium enriched region which suggests necrotic tissue to be responsible for the tracer uptake.
- PET could be used to monitor biodistribution of any potential BNCT compound as long as it could be labeled with a positron emitting isotope. The non-invasive procedure would be widely applicable and, moreover, bypass the need of tissue biopsy or animal sacrifice.
- Control experiments using agents such as radioactive thallium or, possibly, a perfusion agent should demonstrate input functions and validate the proposed kinetic model.

The usefulness of PET is likely to improve due to the current development of research instruments giving a resolution of 2 mm. High resolution PET would also allow a more effective evaluation of potential BNCT compounds in animals, would, therefore, improve modeling, and would help in calculating local dose for therapy planing.

The presentations also outlined the use of PET to monitor copper-64 labeled boron-10 labeled porphyrins (BOPP) that has been proposed for use in BNCT. It is as yet uncertain which of the various boron-10 labeled porphyrin analogues may eventually find acceptance for clinical trials since animal studies have indicated toxicity of some porphyrins and apparently also their accumulation in the arterial membrane.

The participants emphasized the great potential of dynamic PET for BNCT in delivering *in vivo* biokinetic data including sequential, time-dependent ratios of boron concentrations in various body sites such as in tumor, normal brain, and circulating blood. Much remains to be done in acquiring sufficient data to validate the technique.

Additional Targets for BNCT

a) Malignant tumors

The issue of using BNCT to treat malignancies other than glioblastoma multiforme and malignant melanoma was discussed at length. The participants agreed that many tumor types in a variety of locations are good candidates for treatment with BNCT. These tumors fail to respond to conventional therapies, are universally fatal and until late in their evolution are generally well localized. They are likely treatable with localized high LET irradiation, especially that in BNCT. Also, BNCT of these tumors may be easier and more successful that of glioblastoma multiforme. Unfortunately, the current resources for evaluating BNCT for tumors other than those presently in clinical trials are limited. In addition, available resources should be concentrated on completing the current clinical trials in order to reach statistically significant data on safety and , if possible, efficacy. The likelihood of completing current clinical trials would suffer from a serious expansion of clinical and laboratory BNCT protocols. Nonetheless, laboratory research and biodistribution studies using appropriate tumor models should be encouraged with the prospect of a potential expansion into clinical application.

On the other hand, the evaluation of BNCT as a boost in fast neutron therapy of selected tumors should continue and evolve into multicenter trials supported by the National Cancer Institute of the National Institutes of Health. Fast neutron therapy has shown significant efficacy in selected tumor classes, and data on dosimetry and toxicity are available. Therefore, addition of BNCT as boost, as discussed above, should result in easily interpretable results and be acceptable to the radiation therapy community. Furthermore, such a clinical trial would provide important information helping research in BNCT. As experience with BNCT accumulates, as neutron beam quality improves and as better and more selective compounds for neutron capture reactions become available, it is likely that tumors other than glioblastoma multiforme and malignant melanoma will be identified for neutron capture therapy.

b) Non-malignant diseases

Two non-malignant diseases have been suggested as potential candidates for treatment by BNCT. Coronary artery restenosis following balloon angioplasty may benefit from BNCT. Secondly, chronic joint diseases such as rheumatoid arthritis or degenerative arthritis have been successfully treated with intraarticular injections of radionuclides such as yttrium-90, a beta emitter; BNCT for synovectomy poses a lower risk than radionuclide-synovectomy. Moreover, because of the high LET irradiation BNCT may be more effective.

Rheumatoid arthritis (RA) is an autoimmune disease characterized by recurrent swollen, inflamed and painful joints. It afflicts 1–2 % of the US population. Since the cause of RA is unknown, patients are treated symptomatically. Anti-inflammatory drugs are effective in approximately 90% of all patients. In the remaining 10% patients, the inflammation in one or more joints will not respond to drugs and a more severe approach is taken. In the USA, the only option is surgical synovectomy, a costly and painful procedure followed by extensive physical therapy and rehabilitation. Symptomatic relief lasts roughly 2–5 years since the cause of RA has not been addressed.

Radionuclide synovectomy using beta-particle emitters injected directly into the joint is routinely used in Europe and elsewhere and gives about the same symptomatic relief, for the same fraction of patients, for the same length of time, as surgery. Radionuclide-synovectomy is less costly, less painful and requires no rehabilitation time relative to surgery. It is, however, not approved for routine clinical use in the USA due to concerns regarding healthy tissue irradiation caused by leakage of the beta-emitter away from the joint.

Boron Neutron Capture Synovectomy (BNCS) is proposed as a way to carry out radiation synovectomy without the concern regarding leakage of a radioactive substance. A boron-10 labeled compound injected into the joint space would be followed by local irradiation with a beam of low-energy neutrons.

To-date, extensive investigation of BNCS has been carried out. This work involves both the testing of boron-10 labeled compounds also *in vivo* and the design and construction of accelerator-based neutron beams specifically for this purpose. The two compounds checked so far are expected to affect specifically different synovial regions. This approach arose out of the uncertainty as to which cells, if not all, need to be ablated. The energy deposition from the boron neutron capture reaction ranges only over about 15 μm, whereas surgical and radionuclide synovectomy using β-emitters seek to destroy the entire synovium consisting of the subsynovium and the cells lining the joint cavity. It may, however, be necessary to destroy only the phagocytic and enzyme-releasing cavity lining cells. The compounds investigated are potassium-duodecaborane, $K_2B_{12}H_{12}$, expected to pass through the entire synovial membrane, and boron metal particulate, taken up by the phagocytic lining cells only.

Compounds are evaluated, first, using samples of human arthritic synovium taken from the surgical operating room. Co-incubation with the boron compound for a given period of time precedes boron-10 assessment by prompt gamma neutron activation analysis at the MIT reactor. If a compound appears promising following uptake and washout studies in the tissue biopsies, it is then evaluated *in vivo*. For this, the antigen-induced arthritis model in the rabbit is used. Following intra-articular injection of the compound, the animal is sacrificed at a given time for examination of tissues such as synovium, cartilage, ligaments, bone, various organs and fluids for boron-10 uptake analysis.

Results to date are encouraging. In the rabbit model, boron-10 uptake in the synovium at about 20 min after intra-articular injection of $K_2B_{12}H_{12}$ ranges from 265 to 950 ppm. This concentration level has fallen to 30–50 ppm boron-10 at one hour later. While this level is more than sufficient to evaluate the efficacy of BNCS in the animal model, the rapid wash out of the synovium may limit the ultimate use in clinical medicine unless injection and neutron irradiation can be timed closely. Better compounds should be developed and evaluated for eventual clinical trials.

A neutron beam for BNCS has been prepared and installed at MIT's Laboratory for Accelerator Beam Applications (LABA). The corresponding D_2O/graphite assembly will allow rabbit knee treatments in 4-13 minutes for a 1 mAmp proton beam, based on the uptake levels of $K_2B_{12}H_{12}$ already observed in this animal model. BNCS of the relatively large human knee, however, will demand more time and an increase in the proton current, which is feasible. In view of having prepared the radiation facility further efforts are focussed on developing and testing of suitable boron compounds for human use.

It is hoped that approval may be obtained from the FDA to study biodistribution and begin clinical trials in three years. The substantial animal work and beam design already accomplished should aid compound development for human use.

Follow-up and Patient Selection

Reporting of clinical trials depends on the study design as determined by its objectives. A detailed description of the requirements of phase I through phase III clinical protocol design is beyond the scope of this report. The current phase I clinical trials of BNCT at the Brookhaven National Laboratory and the Massachusetts Institute of Technology with the Harvard University Medical School lead the way. These trials have been reported at this Workshop and provide excellent examples of how preliminary data are collected to answer the basic question regarding BNCT:

—Can this therapy be delivered safely?

Conventional study design for BNCT is complicated by the binary system involved; a pharmaceutical grade boron-containing chemical compound is used together with a beam of epithermal-thermal neutrons to induce the boron neutron capture reaction in the selected target tissue. In consequence, some ambiguity exists as to what constitutes a phase I and phase II trial. To establish toxicities in a phase I trial and tumor response in a phase II trial, requires an escalation both of the amount of injected boron compound and an escalation of the neutron radiation dose. Obviously, this essential binary requirement for the therapy trial reflects the problems in using conventional methods to describe BNCT in the trial phases.

Initial testing of BNCT for glioblastma multiforme and malignant melanoma primarily aims at confirming safety, as determined mainly by avoidance of therapy induced CNS complications. Follow-up examinations with neurodiagnostic imaging such as MRI and CT allow the recognition of structural changes associated with functional neurological complications. These may be related to cerebral edema, and also tissue degeneration in the form of necrosis. So far, autopsy data is essential to differentiate tumor progression from radiation necrosis even if functional imaging, for example with PET, may give the same diagnosis *in vivo*, i.e., non-invasively, either prophylactically as part of the follow-up after BNCT or specifically at the time of appearance of symptoms. The most obvious endpoint for these studies is survival, which can be compared to historical or case matched controls.

The clinical trials with suboptimal local radiation doses to the tumor, as they were presented at this Workshop, justify the statement that BNCT according to the current protocols appears to be no more harmful than conventional radiation therapy, although its present efficacy is in the same range as that of conventional therapy. Also, the convenience of a single treatment with BNCT helps justify the recommendation to continue the running trials and prepare for optimization of local dose delivery to the tumor. When BNCT trials demonstrate efficacy beyond that of conventional aggressive therapy, controlled phase III trials will be necessary to prove its benefit and should be an interinstitutional effort funded by the National Institutes of Health.

The participants acknowledged 12.5 Gy to be the limit of tolerance of the normal brain exposed to BNCT. This makes escalation of compound administration more crucial than that of radiation dose in order to optimize doses to the tumors. Thus, more than 350 mg of BPA per kg body weight may eventually be given per treatment session. The foreseen installation of fission plates at the MIT reactor and the BMMR will enhance the neutron fluence, shorten the irradiation times and eases the expected fractionation modality of exposure. In order to avoid problems in evaluation of patient data, BNCT for glioblastoma multiforme should be restricted to histologically defined tumors of given size ranging from 60 to 70 cm³ (including surrounding edema) and not causing midline shift.

The patients should be fully informed of the results of the current clinical trials, with emphasis on the uncertain outcome of the trials and the questionable benefits of BNCT over other aggressive conventional therapies.

Clinical follow-up of all treated patients in close association with the primary physicians is paramount and should be formalized for better data evaluation. The initiation of regular consultations between the Brookhaven and the Boston groups should allow for easy adjustment of protocols and evaluation of data with the hope of better patient referral. Interinstitutional cooperation at the experimental, preclinical and clinical levels should help the assessment and completion of the present trials. Clinical trials at other centers in the USA can be encouraged when the on-going trials indicate efficacy beyond the present level. Studies aimed at various novel approaches to compound delivery such as the combined administration of different boron-10 labeled compounds in conjunction with disruption of the BBB, as discussed at this Workshop, should be continued to maturity for eventual clinical application. A program to obtain clinical pharmacokinetic, biodistribution, and toxicity data to justify seeking FDA approval of intracarotid injection of BSH is underway.

Summarizing Statements

- BNCT in the present protocols is safe and well tolerated.
- The benefit of BNCT is still uncertain.
- The trial results justify BNCT optimization.
- New compounds providing for neutron capture should be tested interinstitutionally.
- BNCT is promising as adjuvant to fast neutron therapy.
- BNCT may be effective for intractable rheumatoid joint disease in peripheral locations.
- The on-going and future clinical trials need to be a coordinated effort.
- Completion of current clinical phase I trials is urgently needed.

Expected Outcome

- Guidance to DOE in managing the BNCT program.
- Immense benefit to all from the open and critical discussions.

Appendix A: Workshop Agenda

Sunday, November 2:

7:00 PM: Informal discussion of agenda *R.F. Hirsch, L.E. Feinendegen and meeting participants*

9:00 PM Adjournment

Monday, November 3:

8:30 AM Welcome and Introduction *R.F. Hirsch, L.E. Feinendegen*

9:00 AM I. BNCT Expectations and Challenges *O.K. Harling*

10:00 AM II. Presentation of Clinical Data,
- Experience at Harvard Medical School and Massachusetts Institute of Technology *P.M. Busse, R.G. Zamenhof*
- Experience at Brookhaven National Laboratory *A. Chanana, J.A. Coderre, A.Z. Diaz*
- The Japanese experience *G.E. Laramore*

1:30 PM III. Treatment Planning and Dosimetry
- Primary and secondary radiation units in BNCT *R.A. Gahbauer*
- Neutron dose, compound concentration, cell and tissue dose *D.M. Nigg, F.J. Wheeler*
- Normal tissue tolerance *J.A. Coderre, P.R. Gavin*
- Clinically implemented dosimetry *J. Capala, R.G. Zamenhof*
- BNCT as adjuvant to fast neutron therapy *G.L. Laramore*

Tuesday, November 4:

8:30 AM IV. Compounds for BNCT
- Compounds for clinical BNCT *M.F. Hawthorne*
- In-vivo assessment of compound biodistribution *T.F. Budinger, G.W. Kabalka*
- Retargeting of compounds to tumor cells *A. Soloway*
- Optimal delivery of compound *R.F. Barth*

1:30 PM V. Additional Targets for BNCT
- Malignant tumors *J. Boggan*
- Non-malignant diseases *J.C. Yanch*
  VI. Patient Selection and Follow-up

- Optimal criteria for selection of patients *P. Gutin*
- Follow-up protocol, reporting *M. Predos, J.H. Goodman*

Wednesday, November 5:

8:30 AM	VII. Identification of Issues for Breakout Sessions *L.E. Feinendegen*
9:00 AM	VIII. Breakout Sessions
11:30 PM	IX. General Discussion and Recommendations *D. Joel, T.L. Phillips*
12:45 PM	Closing Statements *R.F. Hirsch, L.E. Feinendegen.*
1:00 PM	Workshop Conclusion

Appendix B: Attendees

Dr. Susan A. Autry, *Sacramento, California*

Dr. Rolf F. Barth, *Columbus, Ohio*

Dr. James Boggan, *Sacramento, California*

Dr. Thomas F. Budinger, Berkeley, California

Dr. Paul Busse, *Boston, Massachusetts*

Dr. Jacek Capala, *Upton, New York*

Dr. Arjun Chanana, *Upton, New York*

Dr. Jeffrey A. Coderre, *Upton, New York*

Dr. A. Z. Diaz, *Upton, New York*

Dr. Ludwig E. Feinendegen, *Germantown, Maryland*

Dr. Reinhard A. Gahbauer, *Columbus, Ohio*

Dr. Patrick R. Gavin, *Pullman, Washington*

Dr. Joseph H. Goodman, *Columbus, Ohio*

Dr. Philip Gutin, *New York, New York*

Dr. Otto K. Harling, *Cambridge, Massachusetts*

Dr. M. Frederick Hawthorne, *Los Angeles, California*

Dr. Roland F. Hirsch, *Germantown, Maryland*

Dr. Darrel Joel, *Upton, New York*

Dr. George W. Kabalka, *Knoxville, Tennessee*

Dr. George L. Laramore, *Seattle, Washington*

Dr. Hungyuan Liu, *McClellan AFB, California*

Dr. Ruimei Ma, *Upton, New York*

Dr. Michael E. Miner, *Columbus, Ohio*

Dr. Trent Nichols, *Knoxville, Tennessee*

Dr. David W. Nigg, *Idaho Falls, Idaho*

Dr. Theodore L. Phillips, *San Francisco, California*

Dr. Richard C. Reba, *Chicago, Illinois*

Dr. Guido Solares, *Boston, Massachusetts*

Dr. Albert Soloway, *Columbus, Ohio*

Dr. Prem C. Srivastava, *Germantown, Maryland*

Dr. Thomas A. Strike, *Bethesda, Maryland*

Dr. Scott E. Taylor, *Berkeley, California*

Mr. Floyd J. Wheeler, *Idaho Falls, Idaho*

Dr. Richard Wiersema, *San Diego, California*

Dr. Jacqueline C. Yanch, *Cambridge, Massachusetts*

Dr. Robert G. Zamenhof, *Boston, Massachusetts*

Bibliography

1. Locher, G. L. *Am. J. Roentgenol. Radium Ther.* **1936**, *36*, 1–13.

2. (a) Mughabghab, S. F.; Divadeenam, M.; Holden, N. E. *Thermal Neutron Cross Sections*, Academic Press: New York, **1981**. (b) Friedlander, G.; Kennedy, J. W.; Macias, E. S.; Miller, J. M. *Nuclear and Radiochemistry*, 3rd edn.; John Wiley & Sons: New York, **1981**. (c) Leinweber, G.; Barry, D. P.; Trbovich, M. J.; Burke, J. A.; Drindak, N. J.; Knox, H. D.; Ballad, R. V.; Block, R. C.; Danon, Y.; Severnyak L. I. *Nucl. Sci. Eng.* **2006**, *154*, 261–279.

3. Auger, P. *J. Phys.* **1925**, *65*, 205–208.

4. Feinendegen, L. E. *Radiat. Environm. Biophys.* **1975**, *12*, 85–99.

5. Shi, J. L. A.; Brugger, R. M. *Med. Phys.* **1992**, *19*, 733–744.

6. Stepanek, J. In *Advances in Neutron Capture Therapy*, Proceedings of the Seventh International Symposium on Neutron Capture Therapy for Cancer, Zürich, Switzerland, Sept 4–7, 1996.

7. Larsson, B.; Crawford, J.; Weinreich, R., Eds.; Elsevier: Amsterdam, **1997**, Vol. II, pp. 425–429.

8. Allen, B. J.; McGregor, B. J.; Martin, R. F. *Strahlenther. Onkol.* **1989**, *165*, 156–157.

9. Mark, S.; Orion, I.; Laster, B. H.; Shani, G. In Proceedings of the Ninth International Symposium on Neutron Capture Therapy for Cancer, Osaka, Japan, Oct 2–6, 2000; Utsumi, H.; Ono, K.; Kanda, K., Eds.; Research Reactor Institute, Kyoto University: Kyoto, Japan, **2000**, pp. 1411–1415.

10. For summaries see: Various authors in: (a) *Comprehensive Organometallic Chemistry I (COMC-I)*, Vol. 1; Abel, E. W.; Stone, F. G. A.; Wilkinson, G., Eds.; Pergamon: Oxford, 1982. (b) *COMC II* Vol. 1; Abel, E. W.; Stone, F. G. A.; Wilkinson, G., Eds.; Pergamon: Oxford, 1995. (c) *COMC III* Vol. 3; Crabtree, R. H.; Mingos, D. M. P.; Eds.; Elsevier: Oxford, 2006. (d) Caravan, P.; Ellison, J. J.; McMurry, T. J.; Lauffer, R. B. *Chem. Rev.* **1999**, *99*, 2293–2352. (e) Soloway, A. H.; Tjarks, W.; Barnum, B. A.; Feng-Guang Rong, F.-G.; Barth, R. F.; Codogni, I. M.; Wilson, J. G. *Chem. Rev.* **1998**, *98*, 1515–1562. (f) Hawthorne, M. F. *Angew. Chem., Int. Ed. Engl.* **1993**, *32*, 950–984. (g) Rana, G.; Vyakaranam, K.; Maguire, J. A.; Hosmane, N. S. In *Metallotherapeutic Drugs and Metal-Based Diagnostic Agents: The Use of Metals in Medicine*; Gielen, M.; Tiekink, E. Eds.; John Wiley & Sons: New York, **2005**, Chap. 2, pp. 19–49.

11. (a) Salt, C.; Lennox, A. J.; Takagaki, M.; Maguire, J. A.; Hosmane, N. S. *Izv. Akad. Nauk, Ser. Khim.* **2004**, 1785–1812; *Russ. Chem. Bull. Int. Ed.* **2004**, *53*, 1871–1888. (b) Fairchild, R.G.; Bond, V. P. *Int. J. Radiat. Oncol. Biol. Phys.* **1985**, *11*, 831.

12. Guner, A. UA Scientists Discover Cancer Stimulators in Liquid Tumors. *Arizona Daily Wildcat*, Mar 29, 2001, http://wc.arizona.edu/papers/94/125/01_4_m.html.

13. Chabner, D. Cancer Medicine (Oncology); *The Language of Medicine*; Sanders: Missouria, **2007**, pp. 770–771 and 769–799.

14. What is a Gene Mutation and How Do Mutations Occur? *Genetics Home Reference,* Sept 6, 2010, http://ghr.nlm.nih.gov/handbook/mutationsanddisorders/genemutation.

15. Radiation Exposure and Cancer. *American Cancer Society*, http://www.cancer.org/Cancer/CancerCauses/OtherCarcinogens/MedicalTreatments/radiation-exposure-and-cancer.

16. Cancer Causing Chemicals. *Super Oxygen Colon Cleanser*, http://www.oxymega.com/cancer.html.

17. James Ou, J.; Benedict Yen, T. S. *Human Oncogenic Viruses*; World Scientific Publishing Company: Singapore, Dec 9, 2009.

18. Oncovirus. *BlurbWire*, http://www.blurbwire.com/topics/Oncovirus.

19. Hepatitis B Vaccine. *Centers for Disease Control and Prevention*, July 18, 2007, http://www.cdc.gov/vaccines/vpd-vac/hepb/default.htm.

20. Learn about Gardasil. *GARDASIL,* http://www.gardasil.com/what-is-gardasil/index.html.

21. Young, Jr., J. L.; Ward, K. C.; Gloeckler Ries, L. A. Cancers of Rare Cites. *SEER Survival Monograph,* http://seer.cancer.gov/publications/survival/surv_rare_cancers.pdf.

22. Pathologist Job Description, Career as a Pathologist, Salary, Employment — Definition and Nature of the Work, Education and Training Requirements, Getting the Job. *Student University,* http://careers.stateuniversity.com/pages/412/Pathologist.html.

23. Luna, M. A. Pathology of Neck Dissections. *United States of Canadian Pathology,* http://www.pathologyportal.org/95th/pdf/companion14h3.pdf.

24. What are Margins in Relation to Breast Cancer? *Breast Cancer Treatment Information and Pictures,* Apr 23, 2007, http://www.breastcancer.org/questions/margins.jsp.

25. Immunohistochemistry. *Histochem,* http://www.histochem.net.

26. Tamoxifen. *Breast Cancer Treatment Information and Pictures,* Oct 27, 2009, http://www.breastcancer.org/treatment/hormonal/serms/tamoxifen.jsp.

27. Fluorescence in Situ Hybridization (FISH). *JRANK,* http://science.jrank.org/pages/2775/Fluorescence-in-Situ-Hybridization-FISH.html.

28. Tumor Grade. *National Cancer Institute,* May 19, 2004, http://www.cancer.gov/cancertopics/factsheet/Detection/tumor-grade.

29. CancerGuide: Understanding Cancer Types and Staging. *CancerGuide,* http://cancerguide.org/basic.html.

30. *International Federation of Gynecology and Obstetrics,* http://www.figo.org.

31. Staging. *American Cancer Society,* http://www.cancer.org/Treatment/UnderstandingYourDiagnosis/staging.

32 What is Laproscopic Surgery? *Center of Pancreatic and Biliary Diseases,* http://www.surgery.usc.edu/divisions/tumor/pancreasdiseases/web%20pages/laparoscopic%20surgery/WHAT%20IS%20LAP%20SURGERY.html.

33. da Vinci ... Changing The Experience of Surgery. *Da Vinci Surgery,* http://www.davincisurgery.com.

34. Carson-DeWitt, R. Radiofrequency Ablation. *The Doctors of USC,* http://www.doctorsofusc.com/condition/document/146702.

35. Brachytherapy. *RadiologyInfo,* http://www.radiologyinfo.org/en/info.cfm?pg=brachy.

36. Conformal Radiotherapy, 3D Conformal Radiotherapy (3DCRT). *CancerHelp UK,* http://www.cancerhelp.org.uk/about-cancer/cancer-questions/conformal-radiotherapy-3d-conformal-radiotherapy.

37. Gamma Knife. *Mayo Clinic*, http://www.mayoclinic.org/stereotactic-radiosurgery/.

38. What is the CyberKnife? *Cyberknife Robotic Radio Surgery System*, http://www.cyberknife.com/cyberknife-overview/what-cyberknife.aspx.

39. The Voice of the Proton Community. *The National Association for Proton Therapy (NAPT)*, http://www.proton-therapy.org/.

40. Chemotherapy. *Breast Cancer Treatment Information and Pictures*, http://www.breastcancer.org/treatment/chemotherapy.

41. Mechanisms of Alkylating Agents. *Emory University*, http://www.cancerquest.org/index.cfm?page=486.

42. Antimetabolites. *Emory University*, http://www.cancerquest.org/index.cfm?page=429.

43. Mitotic Inhibitors. Drugs.com, http://www.drugs.com/drug-class/mitotic-inhibitors.html.

44. Daniels, M. P. Colchicine Inhibitition of Nerve Fiber Formation *In Vitro*. *J. Cell Biol*. Jan **1972**, *53*(1), http://www.ncbi.nlm.nih.gov/pmc/articles/PMC2108705/.

45. Definition of Topoisomerase Inhibitor. *National Cancer Institute*, http://www.cancer.gov/dictionary/?CdrID=46665.

46. Anthracyclines. *Toxipedia*, http://toxipedia.org/display/toxipedia/Anthracyclines.

47. Hematopoietic Stem Cells. *The National Institutes of Health Resource for Stem Cell Research*, http://stemcells.nih.gov/info/scireport/chapter5.asp.

48. Autologous Bone Marrow Transplant. *Lymphomation.org*, http://www.lymphomation.org.

49. Graft-versus-host Disease. *Medline Plus*, http://www.nlm.nih.gov/medlineplus/ency/article/001309.htm.

50. Hormonal Therapy. *BreastCancer.org*, Oct 23, 2009, http://www.breastcancer.org/treatment/hormonal/.

51. Prostate Cancer Treatment. *National Cancer Institute*, Nov 18, 2009, http://www.cancer.gov/cancertopics/pdq/treatment/prostate/Patient.

52. Aromatase Inhibitors. *BreastCancer.org*, July 21, 2010, http://www.breastcancer.org/treatment/hormonal/aromatase_inhibitors/.

53. Monoclonal Antibodies. *Biology — Davidson*, http://www.bio.davidson.edu/Courses/molbio/MolStudents/01rakarnik/mab.html.

54. About Rituxan. *RITUXAN Treatment for Certain Types of NHL*, http://www. rituxan.com/ra/patient/about/index.xhtml.

55. Angiogenesis Inhibitors Therapy. *National Cancer Institute*, Nov 2008, http://www.cancer.gov/cancertopics/factsheet/Therapy/angiogenesis-inhibitors.

56. Bevacizumab Injection. *National Center for Biotechnology Information*, Dec 2008, http://www.nlm.nih.gov/medlineplus/druginfo/meds/a607001.html.

57. Cancer Growth Blockers (Tyrosine Kinase Inhibitors and Proteasome Inhibitors). *Cancer Research UK*, http://www.cancerhelp.org.uk/about-cancer/treatment/biological/types/cancer-growth-blockers.

58. Faivre, S.; Kroemer, G.; Raymond, E. Current Development of MTOR Inhibitors as Anticancer Agents. *Nat. Rev. Drug Discov.* **2006**, *5*, 671–688.

59. Atkins, M.; Yasothan, U.; Kirkpatrick, P. Everolimus. *Nat. Rev. Drug Discov.* **2009**, *8*, 535–536, http://pubchem.ncbi.nlm.nih.gov/summary/summary.cgi?sid=17396881#Subinfo.

60. Immunotherapy. *American Cancer Society*, http://www.cancer.org/Treatment/TreatmentsandSideEffects/TreatmentTypes/Immunotherapy/index.

61. Park, J.; Benz, C. Immunotherapy Cancer Treatment. *Cancer Supportive Care Programs*, http://www.cancersupportivecare.com/immunotherapy.html.

62. What is a Clinical Trial? *National Cancer Institute*, http://www.cancer.gov/clinicaltrials/education/what-is-a-clinical-trial.

63. Eakin, E.; Youlden, D.; Baade, P.; Lawler, S.; Reeves, M.; Heyworth, J. Health Status of Long-term Cancer Survivors: Results from an Australian Population-Based Sample. *Cancer Epidemiology, Biomarkers & Prevention* Oct **2006**, *15*, 1969.

64. The Childhood Cancer Survivor Study: An Overview. *National Cancer Institute*, Oct 1, 2007, http://www.cancer.gov/cancertopics/coping/ccss.

65. Cancer Prevention: 7 Steps to Reduce Your Risk. *Mayo Clinic*, http://www.mayoclinic.com/health/cancer-prevention/ca00024.

66. Quitting Smoking: Why to Quit and How to Get Help. *National Cancer Institute*, Apr 17, 2007, http://www.cancer.gov/cancertopics/factsheet/Tobacco/cessation.

67. Laws, E. R.; Shaffrey, M. E. *Int. J. Dev. Neurosci.* **1999**, *17*, 413.

68. Barth, R. F.; Coderre, J. A.; Vicente, M. G. H.; Blue, T. E. *Clin. Cancer Res.* **2005**, *11*, 3987.

69. Kruger, P. G. *Proc. Natl. Acad. Sci.* **1940**, *26*, 181–192.

70. (a) Sweet, W. H.; Javid, M. *J. Neurosurg.* **1952**, *9*, 200–209. (b) Farr, L. E.; Sweet, W. H.; Robertson, J. S.; Foster, S. G.; Locksley, H. B.; Sutherland, D. L.; Mendelsohn, M. L.; Stickey, E. E. *Am. J. Roentgenol.* **1954**, *71*, 279–291. (c) Godwin, J. T.; Farr, L. E.; Sweet, W. H.; Robertson, J. S. *Cancer* **1955**, *8*, 601–615. (d) Asbury, A. K.; Ojean, R. G.; Nielsen, S. L.; Sweet, W. H. *J. Neuropathol. Exp. Neurol.* **1972**, *31*, 278–303. (e) Slatkin, D. N. *Brain* **1991**, *114*, 1609–1629.

71. Sweet, W. H. *J. Neuro-Oncol.* **1997**, *33*, 19.

72. Kouri, M.; Kankaanranta, L.; Seppälä, T.; Tervo, L.; Rasilainen, M.; Minn, H.; Eskola, O.; Vähätalo, J.; Paetau, A.; Savolainen, S.; Auterinen, I.; Jääskeläinen, J.; Joensuu, H. *Radiother. Oncol.* **2004**, *72*, 83.

73. (a) Nakagawa, Y.; Hatanaka, H. *J. Neuro-Oncol.* **1997**, *33*, 105. (b) Mishima, Y. In *Cancer Neutron Capture Therapy*; Mishima, Y., Ed., Plenum Press: New York, **1996**, p. 1.

74. Sauerwein, W.; Hideghéty, K.; Gabel, D.; Moss, R. L. *Nucl. News* **1998**, *41*, 54.

75. (a) Tjarks, W.; Anisuzzaman, A. K. M.; Liu, L. *et al. J. Med. Chem.* **1992**, *35*, 1628. (b) Mauer, J. L.; Serino, A. J.; Hawthorne, M. F. *Organometallics* **1988**, *7*, 2519.

76. Hariharan, J. R.; Wyzlic, I. M.; Soloway, A. H. *Polyhedron* **1995**, *14*, 823.

77. (a) Anisuzzman, A. K. M.; Alan, F.; Soloway, A. H. *Polyhedron* **1990**, *9*, 891. (b) Goudgaon, N. M.; Kattan, F. E. G.; Liotta, D. C.; Sxhinazi, R. F. *Nucleosides Nucleotides* **1994**, *13*, 849. (c) Byun, Y.; Narayanasamy, S.; Johnsamuel, J. *et al. Anti-Cancer Agents Med. Chem.* **2006**, *6*, 127.

78. (a) Kahl, S. B.; Koo, M. S. *J. Chem. Soc. Chem. Commun.* **1990**, *24*, 1769. (b) Chen, C. J.; Kane, R. R.; Primus, F. J.; Szalai, G.; Hawthorne, M. F.; Shively, J. E. *Bioconjugate Chem.* **1994**, *5*, 557.

79. (a) Katan, F. E. G.; Lesnikowski, Z. J.; Yao, S.; Tanious, F.; Wilson, W. D.; Schinazi, R. F. *J. Am. Chem. Soc.* **1994**, *116*, 7494. (b) Renner, M. W.; Miura, M.; Easson, M. W.; Vicente, M. G. H. *Anti-Cancer Agents Med. Chem.* **2006**, *6*, 145. (c) Ol'shevskaya, V. A.; Guil'malieva, M. A.; Zaitsev, A. V. *et al.* Boron-Containing Porphyrins: A New Synthetic Approach. *Boron Chemistry at the Beginning of the 21st Century*, Proceeding of

International Conference on Chemistry, Moscow, Russian Federation, 2002; pp. 327–333.

80. Shelly, K.; Feakes, D. A.; Hawthorne, M. F.; Schmidt, P. G.; Krisch, T. A.; Bauer, W. F. *Proc. Natl. Acad. Sci. USA* **1992**, *89*, 9039.

81. (a) Radel, P. A.; Kahl, S. B. *J. Org. Chem.* **1996**, *61*, 4582. (b) Kabalka, G. W.; Yao, M. L. *Anti-Cancer Agents Med. Chem.* **2006**, *6*, 111.

82. Fukuda, H.; Hiratsuka, J.; Honda, C. *et al. Radiat. Res.* **1994**, *138*, 435.

83. Coderre, J. A.; Elowitz, E. H.; Chadha, M. *et al. J. Neuro-Oncol.* **1997**, *33*, 141.

84. Elowitz, E. H.; Bergland, R. M.; Coderre, J. A.; Joel, D. D.; Chadha, M.; Chanana, A. D. *Neurosurgery* **1998**, *42*, 463.

85. Soloway, A. H.; Hatanaka, H.; Davis, M. A. *J. Med. Chem.* **1967**, *10*, 714.

86. Hatanaka, H.; Nakagawa, Y. *Int. J. Radiat. Oncol. Biol. Phys.* **1994**, *28*, 1061.

87. Mishima, Y., Ed., *Cancer Neutron Capture Therapy.* Plenum Press: New York, **1996**; pp. 1–26.

88. Van Rij, C. M.; Wilhelm, A. J.; Sauerwein, W.; van Loenen, A. *Pharm. World Sci.* **2005**, *27*, 92.

89. (a) Chen, W.; Mehta, S. C.; Lu, D. R. *Adv. Drug. Deliv. Rev.* **1997**, *26*, 231. (b) Gong, W.; Barth, R. F.; Yang, W. *et al. Anti-Cancer Agents Med. Chem.* **2006**, *6*, 167.

90. (a) Partridge, W. M *NeuroRx* **2005**, *2*, 3–14. For a discussion of delivery through the BBB, see various authors in: *NeuroRx* **2005,** *2*(1), 1–161. (b) Hediger, M. A.; Romero, M. F.; Peng, J.-B.; Rolfe, A.; Takanaga, H.; Bruford, E. A. *Eur. J. Physiol.* **2004**, *447*, 465–468. (c) Abbott, N. J.; Rönnbäck, L.; Hansson, E. *Nat. Neurosci.* **2006**, *7*, 41–53. (d) For a recent review, see: various authors. *Neurobiol. Dis.* **2010**, *37*(1), 1–236.

91. Yokoyama, K.; Miyatake, S.-I.; Kajimoto, Y.; Kawabata, S.; Doi, A.; Yoshida, T.; Asano, T.I.; Kirihata, M.; Ono, K.; Kuroiwa, T. *J. Neuro-Oncol.* **2006**, *78*, 227–232.

92. Suzuki, M.; Sakurai, Y.; Nagata, K.; Kinashi, Y.; Masunaga, S.; Ono, K.; Maruhashi, A.; Kato, I.; Fuwa, N.: Hiratsuka, J.; Imahori, Y. *Int. J. Radiat. Oncol. Biol. Phys.* **2006**, *66*, 1523–1527.

93. Kageji, T.; Nakagawa, Y.; Kitamura, K.; Matsumoto, K.; Hatanaka, H. *J. Neuro-Oncol.* **1997**, *33*, 117–130.

94. Yang, W.; Barth, R. F.; Rotaru, J. H.; Moeschberger, M. L.; Joel. D. D.; Nawrocky, M. M.; Goodman, J. H. *J. Neuro-Oncol.* **1997**, *33*, 59–70.

95. Barth, R. F.; Yang, W.; Coderre, J. A. *J. Neuro-Oncol.* **2003**, *62*, 61–74.

96. Pardridge, W. M. *Brain Drug Targeting: The Future of Brain Development.* Cambridge University Press: New York, **2001**.

97. Kurihara, A.; Deguchi, Y.; Pardridge, W. M. *Bioconjugate Chem.* **1999**, *10*, 502–511.

98. (a) Shapiro, W. R.; Voorhies, R. M.; Hiesiger, E. M.; Sher, P. B.; Basler, G. A.; Lipschutz, L. E. *Cancer Res.* **1988**, *48*, 694. (b) Elkassabany, N. M.; Bhatia, J.; Deogaonkar, A.; Barnett, G. H.; Lotto, M.; Maurtua, M.; Ebrahim, Z.; Shubert, A.; Ference, S.; Farag, E. *J. Neurosurg. Anesth.* **2008**, *20*, 45–48.

99. Bickel, U.; Yoshikawa, T.; Pardridge, W. M. *Adv. Drug Deliv. Rev.* **2001**, *46*, 247–279.

100. (a) Yang, W.; Barth, R. F.; Adams, D. M.; Ciesielski, M. J.; Fenstermaker, R. A.; Shukla, S.; Tjarks, W.; Caligiuri, M. A. *Cancer Res.* **2002**, *62*, 6552–6558. (b) Voges, J.; Rezka, R.; Grossmann, A.; Dittmar, C.; Richter, R.; Garlip, G.; Kracht, L.; Coenen, H. H.; Strum, V.; Weinhard, K.; Heiss, W.-D.; Jacobs, A. H. *Ann. Neurol.* **2003**, *54*, 479–487.

101. Farr, L. E.; Sweet, W. H.; Robertson, J. S.; Foster, C. G.; Locksley, H. B.; Sutherland, D. L.; Mendelsohn, M. L.; Stickley, E. E. *Am. J. Roentgenol. Radium Ther. Nucl. Med.* **1954**, *71*, 279.

102. Farr, L. E.; Sweet, W. H.; Locksley, H. B.; Robertson, J. S. *Trans. Am. Neurol. Assoc.* **1954**, *79*, 110.

103. Locksley, H. B.; Farr, L. E. *J. Pharmacol. Exp. Ther.* **1955**, *114*, 4.

104. Godwin, J. T.; Farr, L. E.; Sweet, W. H.; Robertson, J. S. *Cancer* **1955**, *8*, 601.

105. Hawthorne, M. F.; Pitochelli, A. R. *J. Am. Chem. Soc.* **1959**, *81*, 5519.

106. (a) Miller, H. C.; Miller, N. E.; Muetterties, E. L. *J. Am. Chem. Soc.* **1963**, *85*, 3885. (b) Ellis, I. A.; Gains, D. F.; Schaeffer, R. *J. Am. Chem. Soc.* **1963**, *85*, 3885.

107. Soloway, A. H. *Science* **1958**, *128*, 1572.

108. Sweet, W. H. *J. Neuro-Oncol.* **1997**, *33*, 19.

109. (a) Ishihashi, M.; Nakanishi, T.; Mishima, Y. *J. Invest. Dermatol.* **1982**, *78*, 215. (b) Mishima, Y.; Ishihashi, M.; Nakanishi, T.; Tsuji, M.; Ueda, M.; Nakagawa, T. Proceeding of First International Symposium Neutron Capture Therapy, Brookhaven National Laboratory, BNL 1983, BNL 51730, p. 355. (c) Mishiina, Y.; Honda, C.; Ichihashi, M.; Obara, H.; Hiratsuka, J.; Yoshino, K. *Lancet* **1989**, *2*, 388–389.

110. (a) Coderre, J. A.; Joel, D. D.; Micca, P. L.; Nawrocky, M. M.; Slatkin, D. N. *Radiat. Res.* **1992**, *129*, 290. (b) Coderre, J. A.; Glass, J. D.; Fairchild, R. G.; Micca, P. L.; Fand, I.; Joel, D. D. *Cancer Res.* **1990**, *50*, 138.

111. Soloway, A. H.; Hatanaka, H.; Davis, M. A. *J. Med. Chem.* **1967**, *10*, 714–717.

112. Nakagawa, Y.; Hatanaka, H. *J. Neuro-Oncol.* **1997**, *33*, 105–115.

113. Slatkin, D. N.; Joel, D. D.; Fairchilf, R. G.; Micca, P. L.; Nawrocky, M. M.; Laster, B. H.; Coderre, J. A.; Finkel, G. C.; Poletti, C. E.; Sweet, W. H. *Basic Life Sci.* **1989**, *50*, 179–191.

114. (a) Gilbert, B.; Perfetti, L.; Fauchoux, J.; Baudat, P.-A.; Andres, R.; Neumann, M.; Steen, S.; Gabel, D.; Mercanti, D.; Ciotti, M. T.; Perfette, P.; Margaritondo, G.; De Stasio, G. *Phys. Rev. E* **2000**, *62*, 1110–1118. (b) Gibson, C. R.; Staubus, A. E.; Barth, R. F.; Yang, W.; Kleinholz, N. M.; Jones, R. B.; Green-Church, K.; Tjarks, W.; Soloway, A. H. *Drug Metab. Dispos.* **2001**, *29*, 1588–1598. (c) Bendel, P.; Wittig, A.; Basilico, F.; Mauri, P. L.; Sauerwein, W. *J. Pharm. Biomed. Anal.* **2010**, *51*, 284–287. (d) Svantesson, E.; Capala, J.; Markides, K. E.; Pettersson, J. *Anal. Chem.* **2002**, *74*, 5358–5363.

115. (a) Barth, R. F.; Rotaru, J. H.; Staubus, A. E.; Soloway, A. H.; Moeschberger, M. L. In *Cancer Neutron Capture Therapy*; Mishima, Y., Ed.; Plenum Press: New York **1996**; p. 769. (b) Yokoyama, K.; Miyatake, S.-L.; Kajimoto, Y.; Kawabata, S.; Doi, A.; Yoshida, T.; Okabe, M.; Kirihata, M.; Ono, K.; Kuroiwa, T. *Radiat. Res.* **2007**, *167*, 102–109. (c) Ono, K.; Masunaga, S.-I.; Kinashi, Y.; Takagaki, M.; Akaboshi, M.; Kobayashi, T.; Eng, D.; Akuta, K. *Int. J. Radiat. Oncol. Biol. Phys.* **1996**, *34*, 1081–1086.

116. (a) Burnham, B. S. *Curr. Med. Chem.* **2005**, *12*, 1995–2010. (b) El-Zaria, M. E.; Genady, A. R.; Gabel, D. *Chem. Eur. J.* **2006**, *12*, 8084–8089. (c) El-Zaria, M. E.; Dörfler, U.; Gabel, D. *J. Med. Chem.* **2002**, *45*, 5817–5819.

117. Pegg, A. E. *Cancer Res.* **1988**, *48*, 759–774.

118. (a) Bergstand, N.; Bohl, E.; Carlsson, J.; Edwards, K.; Ghaneolhosseine, H.; Gedda, L.; Johnsson, M.; Silvander, M.; Sjöberg, S. In *Contemporary Boron Chemistry,* Royal Society of Chemistry: Cambridge, **2000**. (b) Zhou, J.-C.; Cai, J.; Soloway, A. H.; Barth, R. F.; Adams, D. M.; Ji, W.; Tjarks, W. *J. Med. Chem.* **1999**, *42*, 1282–1292.

119. Crossley, E. L.; Ziolkowski, E. J.; Coderre, J. A.; Rendina, L. M. *Mini-Rev. Med. Chem.* **2007**, *7*, 303–313 and references therein.

120. Lee, J.-D.; Lee, Y.-J.; Jeong, H.-J.; Lee, J. S.; Lee, C.-H.; Ko, J.; Kang, S. O. *Organometallics* **2003**, *22*, 445.

121. Malmquist, J. S.; Sjöberg, J. *Inorg. Chem.* **1992**, *31*, 2534.

122. Cai, J.; Soloway, A. H.; Barth, R. F.; Adams, D. M.; Hariharan, J. R.; Wyzlic, I. M.; Radcliffe, K. *J. Med. Chem.* **1997**, *40*, 3887–3896.

123. Zhu, Y. *Bis(aminoalkyl)-dicarbaborane Derived Boron Neutron Capture Therapy Drugs*, **2006**, WO27058630 A1.

124. Zhu, Y.; Zhang, H.; Maguire, J. A.; Hosmane, N. S. *Inorg. Chem. Commun.* **2001**, *4*, 447.

125. Mori, Y.; Suzuki, A.; Yoshino, K.; Kakihana, H. *Pigment Cell Res.* **1989**, *2*, 273.

126. Nemoto, H.; Iwamoto, S.; Nakamura, H.; Yamamoto, Y. *Chem. Lett.* **1993**, *22*, 465.

127. Karnbrock, W.; Musiol,. H. J.; Moroder, L. *Tetrahedron* **1995**, *51*, 1187.

128. (a) Wyzlic, I. M.; Soloway, A. H. *Tetrahedron Lett.* **1992**, *33*, 7489. (b) Adel, P. A.; Kahl, S. B. *J. Org. Chem.* **1996**, *61*, 4582. (c) Prashar, J. K.; Moore, D. E. *J. Chem. Soc. Perkin. Trans.* **1993**, 1051.

129. Pettersson, O.-A.; Olsson, P.; Lindström, P.; Sjöberg, S.; Carlsson, J. *Melanoma Res.* **1993**, *3*, 369.

130. Pettersson, O.-A.; Lindström, P.; Olsson, P.; Carlsson, J.; Sjöberg, S.; Larsson, B. S. In *Advances in Neutron Capture Therapy*; Soloway A. H.; Barth, R. F.; Carpenter, D. E., Eds.; Plenum Press: New York, 1993; p. 629.

131. Spielvogel, B. F.; McPhail, A. T.; Das, M. K.; Hall, I. H. *J. Am. Chem. Soc.* **1980**, *102*, 6343.

132. (a) Scheller, K. H.; Martin, R. B.; Spielvogel, B. F.; McPhail, A. T. *Inorg. Chim. Acta* **1982**, *57*, 227. (b) Kabalke, G. W.; Yao, M. L.; Marepally, S. R.; Chandra, S. *Appl. Radiat. Isot.* **2009**, *67*, S374–S379.

133. Liao, T. K.; Pondrebarac, E. G.; Cheng, C. C. *J. Am. Chem. Soc.* **1964**, *86*, 1869.

134. Schinazi, R. F.; Prusoff, W. H. *Tetrahedron Lett.* **1978**, *50*, 4981.

135. Schinazi, R. F.; Prusoff, W. H. *J. Org. Chem.* **1985**, *50*, 841.

136. Nemoto, H.; Cai, J.; Yamamoto, Y. *J. Chem. Soc., Chem. Commun.* **1994**, 577.

137. Reynolds, R. C.; Trask, T. W.; Sedwick, W. D. *J. Org. Chem.* **1991**, *56*, 2391.

138. Collins, J. M.; Klecker Jr, R. W.; Kelley, J. A.; Roth, J. S.; McCully, C. L.; Balis, F. M.; Poplack, D. G. *J. Pharmcol. Exp. Ther.* **1988**, *245*, 466.

139. Yamamoto, Y. In *Advances in Neutron Capture Therapy*; Larsson, B.; Crawford, J.; Weinreich, R., Eds.; Chemistry and Biology, Elsevier: Amsterdam, **1997**, Vol. II, p. 22.

140. Tjarks, W.; Anisuzzaman, A. K. M.; Liu, L.; Soloway, A. H.; Barth, R. F.; Perkins, D. J.; Adams, D. M. *J. Med. Chem.* **1992**, *35*, 1628.

141. Mangiardi, J. R. In *Brain Tumors. An Encyclopedic Approach*; Kaye, A. H., Laws, E. R., Jr, Eds.; Churchill Livingstone: Edinburgh, **1995**, p. 99.

142. Peymann, T. *Sauerstoff und Schwefel am Undecahydro-closo-dodecaborat(2′) als Nucleophile.* PhD thesis, University of Bremen, Germany, **1995**.

143. Gabel, D.; Harfst, S.; Moller, D.; Ketz, H.; Peymann, T.; Rösler, J. In *Current Topics in the Chemistry of Boron*; Kabalka, G. W., Ed.; The Royal Society of Chemistry: Cambridge, UK, **1994**, p. 161.

144. Hawthorne, M. F. *Pure Appl. Chem.* **1991**, *63*, 327.

145. Peymann, T.; Preusse, D.; Gabel, D. In *Advances in Neutron Capture Therapy*; Larsson, B.; Crawford, J.; Weinreich, R., Eds.; Chemistry and Biology; Elsevier: Amsterdam, The Netherlands, **1997**, Vol. II, p. 35.

146. Lemmen, P.; Werner, B. *Chem. Phys. Lipids* 1992; *62*, 185.

147. Lemmen, P.; Weissfloch, L.; Auberger, T.; Probst, T. *Anti-Cancer Drugs* **1995**, *6*, 744.

148. Lindström, P.; Olsson, P.; Malmquist, J.; Petterson, J.; Lemmen, P.; Sjöberg, S.; Olin, A.; Carlsson, J. *Anti-Cancer Drugs* **1994**, *5*, 43.

149. Selverstone, B.; Sweet, W. H.; Robinson, C. V. *Ann. Surg.* **1949**, *130*, 643.

150. John, K. C.; Kaczmarczyk, A.; Soloway, A. H. *J. Med. Chem.* **1969**, *12*, 54.

151. Bechthold, R. A.; Kaczmarczyk, A. *J. Med. Chem.* **1975**, *18*, 371.

152. Zakharkin, L. I.; Bregadze, V. I.; Okhlobystin, O. Y. *Izv. Akad. Nauk, Ser. Khim.* **1964**, 1539.

153. Degtyarev, A. N.; Godovikov, N. N.; Bregadze, V. I.; Kabachnik, M. L. *Bull. Acad. Sci. USSR* **1974**, *22*, 2314.

154. Fairchild, R. G.; Gabel, D.; Hillman, M.; Watts, K. In *Neutron Capture Therapy*; Hatanaka, H., Ed.; Nishamura Co., Ltd.: Niigata, Japan, 1986; p. 266.

155. Kahl, S. B.; Koo, M.-S. *J. Chem. Soc., Chem. Commun.* **1990**, 1769.

156. Kahl, S. B.; Li, J. *Inorg. Chem.* **1996**, *35*, 3878.

157. Alam, A.; Soloway, A. H.; Bapat, B. V.; Barth, R. F.; Adams, D. M. *Basic Life Sci.* **1989**, *50*, 107.

158. (a) Ozawa, T.; Afzal, J.; Lamborn, K. R.; Bollen, A. W.; Bauer, W. F.; Koo, M. S.; Kahl, S. B.; Deen, D. F. *Int. J. Radiat. Oncol. Biol. Phys.* **2005**, *63*, 247–252. (b) Ozawa, T.; Santos, R. A.; Lamborn, K. R.; Bauer, W. F.; Koo, M. S.; Kahl, S. B.; Deen, D. F. *Mol. Pharm.* **2004**, *1*, 368–374.

159. Renne, M. W.; Miura, M.; Easson, M. W.; Vicente, M. G. H. *Anti-Cancer Agents Med. Chem.* **2006**, *6*, 145–157.

160. (a) Hao, E.; Jensen, T. J.; Courtney, B. H.; Vicente, M. G. *Bioconjugate Chem.* **2005**, *16*, 1495–1502. (b) Sibrian-Vazquez, M.; Hao, E.; Jensen, T. J.; Vicente, M. G. *Bioconjugate Chem.* **2006**, *17*, 928–934.

161. Giuntini, F.; Raoul, Y.; Dei, D.; Municchi, M.; Chiti, G.; Fabris, C.; Colautti, P.; Jori, G.; Roncucci, G. *Tetrahedron Lett.* **2005**, *46*, 2979–2982.

162. Sweet, F. *Steroids* **1981**, *37*, 223.

163. Schneiderova, L.; Strouf, O.; Gruner, B.; Pouzar, V.; Drasar, P.; Hampl, R.; Kimlova, I. *Collect. Czech. Chem. Commun.* **1992**, *57*, 463.

164. Hawthorne, M. F. In *Cancer Neutron Capture Therapy*; Mishima, Y., Ed., Plenum Press: New York, **1996**.

165. Jain, R. K. *Nat. Med.* **2001**, *7*, 987.

166. Santini, J.; Cima, M.; Langer, R. *Nature* **1999**, *397*, 335.

167. Sartor, O.; Dineen, M. K.; Perez-Marreno, R.; Chu, F. M.; Carron, G. J.; Tyler, R. C. *Urology* **2003**, *62*, 319.

168. Yih, T. C.; Al-Fandi, M. *J. Cell Biochem.* **2006**, *97*, 1184.

169. Hobbs, K.; Monsky, W.; Yuan, F.; Roberts, W. G.; Griffith, L.; Torchilin, V. P.; Jain, R. K. *Proc. Natl. Acad. Sci. USA* **1998**, *95*, 4607.

170. Kreuter, J.; Alyyautidin, R. N.; Kharkevich, D. A.; Ivanov, A. A. *Brain. Res.* **1995**, *674*, 171.

171. Chen, W.; Mehta, S. C.; Lu, D. R. *Adv. Drug Deliv. Rev.* **1997**, *26*, 231.

172. Retrieved from *Encyclopædia Britannica* [Online] **2009**.

173. (a) Hawthorne, M. F. *Angew. Chem. Int. Ed. Engl.* **1993**, *32*, 950–984. (b) Shelly, K.; Feakes, D. A.; Hawthorne, M. F.; Schmidt, P. G.; Krisch, T. A.; Bauer, W. F. *Proc. Natl. Acad. Sci., USA* **1992**, *89*, 9039–9043. (c) Feakes, D. A.; Shelly, K.; Knobler, C. B.; Hawthorne, M. F. *Proc. Natl. Acad. Sci., USA* **1994**, *91*, 3029–3033.

174. (a) Gabizon, A.; Price, D. C.; Huberty, J.; Bresalier, R. S.; Papahadjopoulos, D. *Cancer Res.* **1990**, *50*, 6371. (b) Pavanetto, F.; Perugini, P.; Genta, I.; Minoia, C.; Ronchi, A.; Prati, U.; Roveda, L.; Nano, R. *Drug Del.* **2000**, *7*, 97–103.

175. Li, T.; Hamdi, J.; Hawthorne, M. F. *Bioconjugate Chem.* **2006**, *17*, 15–20.

176. Pan, X. Q.; Wang, H.; Shukla, S.; Sekido, M.; Adams, D. M.; Tjarks, W.; Barth, R. F.; Lee, R. J. *Bioconjugate Chem.* **2002**, *13*, 435–442.

177. Torchilin, P. *Nat. Rev. Drug Discov.* **2005**, *4*, 145.

178. Wu, G.; Barth, R. F.; Yang, W.; Lee, R. J.; Tjarks, W.; Backer, M. V.; Backer, J. M. *Anti-Cancer Agents Med. Chem.* **2006**, *6*, 167–184.

179. Hawthorne, M. F.; Feakes, D. A.; Shelly, K. In *Cancer Neutron Capture Therapy*; Mishima, Y., Ed.; Plenum Press: New York, **1996**; p. 27.

180. Hawthorne, M. F.; Shelly, K. *J. Neuro-Oncol.* **1997**, *33*, 53.

181. Nakamura, H.; Miyajima, Y.; Takei, T.; Kasaoka, S.; Maruyama, K. *Chem. Commun.* **2004**, 1910.

182. Ristori, S.; Oberdisse, J.; Grillo, I.; Donati, A.; Spalla, O. *Biophys. J.* **2005**, *88*, 535.

183. Yanagie, H.; Matuyama, K.; Takizawa, T.; Ishida, O.; Matsumoto, T.; Sakurai, Y.; Kobayashi, T.; Shinohara, A.; Skvarc, J.; Ilic, R.; Kuhne, G.; Chiba, M.; Furuya, Y.; Sugiyama, H.; Hisa, T.; Kobayashi, H.; Eriguchi, M. *Biomed. Pharmacother.* **2006**, *60*, 43.

184. Nobuto, H.; Sugita, T.; Kubo, T.; Shimose, S.; Yasunaga, Y.; Murakami, T.; Ochi, M. *Int. J. Cancer* **2004**, *109*, 627–635.

185. Jurgons, R.; Seliger, C.; Hilpert, A.; Trahms, L.; Odenbach, S.; Alexiou, C. *J. Phys.: Condens. Matter* **2006**, *18*, S2893–S2902.

186. Ringsdorf, H. *J. Polym. Sci. Polym. Symp.* **1975**, *51*, 135.

187. Kopecek, J. *Polym. Med. (Wroclaw)* **1977**, *7*, 191.

188. Gillies, E. R.; Frechet, J. M. J. *J. Am. Chem. Soc.* **2002**, *124*, 14137.

189. Seymour, L. W. *Crit. Rev. Ther. Drug Carrier Syst.* **1992**, *9*, 135.

190. (a) Newkome, G. R.; Moorefield, C. N.; Vogtle, F. *Dendritic Macromolecules: Concepts, Synthesis, Perspectives*. VCH: Weinheim, Germany, **1996**. (b) Moorefield, C. N.; Newkome, G. R. *C. R. Chimie* **2003**, *6*, 715–724.

191. Niederhafner, P.; Sebestik, J.; Jezek, J. *J. Peptide Sci.* **2005**, *11*, 757.

192. Gillies, E. R.; Frechet, J. M. J. *Drug Discov. Today* **2005**, *10*, 35.

193. Esfand, R.; Tomalia, D. A. *Drug Discov. Today* **2001**, *6*, 427.

194. Kukowska-Latallo, J. F.; Candido, K. A.; Cao, Z.; Nigavekar, S. S.; Majoros, I. J.; Thomas, T. P.; Balogh, L. P.; Khan, M. K.; Baker Jr, J. R. *Cancer Res.* **2005**, *65*, 5317.

195. Barth, R. F.; Yang, W.; Adams, D. M.; Rotaru, J. H.; Shukla, S.; Sekido, M.; Tjarks, W.; Fenstermaker, R. A.; Ciesielski, M.; Nawrocky, M. M.; Coderre, J. A. *Cancer Res.* **2002**, *62*, 3159.

196. Barth, R. F.; Adams, D. M.; Soloway, A. H.; Alam, F.; Darby, M. V. *Bioconjugate Chem.* **1994**, *5*, 58–66.

197. Shukla, S.; Wu, G.; Chatterjee, M.; Yang, W.; Sekido, M.; Diop, L. A.; Müller, R.; Sudimack, J. J.; Lee, R. J.; Barth, R. F.; Tjarks, W. *Bioconjugate Chem.* **2003**, *14*, 158.

198. Wu, G.; Barth, R. F.; Yang, W.; Chatterjee, M.; Tjarks, W.; Ciesielski, M.; Fenstermaker, R. A. *Bioconjugate Chem.* **2004**, *15*, 185–194.

199. Liu, L.; Bäcklund, L. M.; Nilsson, B. R.; Grandér, D.; Ichimura, K.; Goike, H. M.; Collins, V. P. *J. Mol. Med.* **2005**, *83*, 917–926.

200. Wikstrand, C. J.; Hale, L. P.; Batra, S. K.; Hill, M. L.; Humphrey, P. A.; Kurpad, S. N.; McLendon, R. E.; Moscatello, D.; Pegram, C. N.; Reist, C. J.; Traweek, S. T.; Wong, A. J.; Zalutsky, M. R.; Bigner, D. D. *Cancer Res.* **1995**, *55*, 3140–3148.

201. Yang, W.; Barth, R. F.; Wu, G.; Kawabata, S.; Sferra, T. J.; Bandyopadhyaya, A. K.; Tjarks, W.; Ferketich, A. K.; Moeschberger, M. L.; Binns, P. J.; Riley, K. J.; Coderre, J. A.; Ciesielski, M. J.; Fenstermaker, R. A.; Wikstrand, C. J. *Clin. Cancer Res.* **2006**, *12*, 3792–3802.

202. Yang, W.; Wu, G.; Barth, R. F.; Swindall, M. R.; Bandyopadhyaya, A. K.; Tjarks, W.; Tordoff, K.; Moeschberger, M.; Sferra, T. J.; Binns, P. J.; Riley, K. J.; Ciesielski, M. J.; Fenstermaker, R. A.; Wikstrand, C. J. *Clin. Cancer Res.* **2008**, *14*, 883–891.

203. Yang, W.; Barth, R. F.; Wu, G.; Tjarks, W.; Binns, P.; Riley, K. *Appl. Radiat. Isot.* **2009**, *67*, S328–S331.

204. Gillies, E. R.; Jonsson, T. B.; Fréchet, J. M. J. *J. Am. Chem. Soc.* **2004**, *126*, 11936–11943.

205. (a) Gillies, E. R.; Fréchet, J. M. J. *Bioconjugate Chem.* **2005**, *16*, 361–368. (b) Lee, C. C.; Gillies, E. R.; Fox, M. E.; Guillaudeu, S. J.; Fréchet, J. M. J.; Dy, E. E. *Proc. Natl. Acad. Sci.* **2006**, *103*, 16649–16654.

206. Parrott, M. C.; Marchington, E. B.; Valliant, J. F.; Adronov, A. *J. Am. Chem. Soc.* **2005**, *127*, 12081–12089.

207. Newkome, G. R.; Moorefield, C. N.; Keith, J. M.; Baker, G. R.; Escamilla, G. H. *Angew. Chem. Int. Ed. Engl.* **1994**, *33*, 666–668.

208. Ma, L.; Hamdi, J.; Wong, F.; Hawthorne, M. F. *Inorg. Chem.* **2006**, *45*, 278–284.

209. Iijma, S. *Nature* **1991**, *354*, 56.

210. Various authors in: *Acc. Chem. Res.* **2002**, *35*, 997–1113.

211. (a) Prato, M.; Kostarelos, K.; Bianco, A. *Acc. Chem. Res.* **2008**, *41*, 60–68. (b) Kostarelos, K.; Lacerda, L.; Pastorin, G.; Wu, W.; Wieckowski, S.;

Luangsivilay, J.; Godefroy, S.; Pantarotto, D.; Braind, J.-P.; Muller, S.; Prato, M.; Bianco, A. *Nat. Nanotech.* **2007**, *2*, 108–113.

212. Pantarotto, D.; Briand, J. P.; Prato, M.; Bianco, A. *Chem. Commun.* **2004**, 16.

213. Cherukuri, P.; Bachilo, S. M.; Litovsky, S. H.; Weisman, R. B. *J. Am. Chem. Soc.* **2004**, *126*, 15638.

214. Kam, N. W.; O'Connell, M.; Wisdom, J. A.; Dai, H. *Proc. Natl. Acad. Sci. USA* **2005**, *102*, 11600.

215. Pantarotto, D.; Partidos, C. D.; Graff, R.; Hoebeke, J.; Briand, J.-P.; Prato, M.; Bianc, A. *J. Am. Chem. Soc.* **2003**, *125*, 6160–6164.

216. Kam, N. W. S.; Jessop, T. C.; Wender, P. A.; Dai, H. *J. Am. Chem. Soc.* **2004**, *126*, 6850.

217. Furtado, C. A.; Kim, U. J.; Gutierrez, H. R.; Pan, L.; Dickey, E. C.; Eklund, P. C. *J. Am. Chem. Soc.* **2004**, *126*, 6095.

218. Hu, H.; Ni, Y.; Montana, V.; Haddon, R. C.; Parpura, V. *Nano. Lett.* **2004**, *4*, 507.

219. Fu, K.; Li, H.; Zhou, B.; Kitaygorodskiy, A.; Allard, L. F.; Sun, Y. P. *J. Am. Chem. Soc.* **2004**, *126*, 4669.

220. Georgakilas, V.; Kordatos, K.; Prato, M.; Guldi, D. M.; Holzinger, M.; Hirsch, A. *J. Am. Chem. Soc.* **2002**, *124*, 760.

221. Peng, H.; Alemany, L. B.; Margrave, J. L.; Khabashesku, V. N. *J. Am. Chem. Soc.* **2003**, *125*, 15174.

222. Peng, H.; Reverdy, P.; Khabashesku, V. N.; Margrave, J. L. *Chem. Commun.* **2003**, 362.

223. Huang, W.; Fernando, S.; Lin, Y.; Zhou, B.; Allard, L. F.; Sun, Y. P. *Langmuir* **2003**, *19*, 7084.

224. Dyke, C. A.; Tour, J. M. *Nano. Lett.* **2003**, *3*, 1215.

225. Zhu, W.; Minami, N.; Kazaoui, S.; Kim, Y. *J. Mater. Chem.* **2003**, *13*, 2196.

226. Hu, H.; Zhao, B.; Hamon, M. A.; Kamaras, K.; Itkis, M. E.; Haddon, R. C. *J. Am. Chem. Soc.* **2003**, *125*, 14893.

227. Chen, R. J.; Choi, H. C.; Bangsaruntip, S.; Yenilmez, E.; Tang, X.; Wang, Q.; Chang, Y.-L.; Dai, H. *J. Am. Chem. Soc.* **2004**, *126*, 1563–1658.

228. Zhu, Y. H.; Ang, T. P.; Carpenter, K.; Maguire, J. A.; Hosmane, N. S.; Takagak, M. *J. Am. Chem. Soc.* **2005**, *127*, 9875.

229. Morgan, D. A.; Sloan, J.; Green, M. L. H. *Chem. Commun.* **2002**, *20*, 2442–2443.

230. Xu, T. T.; Zheng, J.-G.; Wu, N.; Nicholls, A. W.; Roth, J. R.; Dikin, D. A.; Ruoff, R. S. *Nano. Lett.* **2004**, *4*, 963–968.

231. Cao, L. M.; Tian, H.; Zhang, Z.; Zhang, X. Y.; Gao, C. X.; Wang, W. K. *Nanotechnology* **2004**, *15*, 139.

232. Ciuparu, D.; Klie, R.; Zhu, Y.; Pfefferle, L. *J. Phys. Chem. B* **2004**, *108*, 3967–3969.

233. (a) Liu. F.; Taain, J.; Bao, L.; Yang, T.; Shen, C.; Lai, X.; Xiao, Z.; Xie, W.; Deng, S.; Chen, J.; She, J.; Xu, N.; Gao, H. *Adv. Mater.* **2008**, *20*, 2609–2615. (b) Wang, X.; Tian, J.; Yang, T.; Bao, L.; Hui, C.; Liu, F.; Shen, C.; Gu, C.; Xu, N.; Gao, H. *Adv. Mater.* **2007**, *19*, 4480–4485. (c) Liu, F.; Shen, C.; Su, Z.; Ding, X.; Deng, S.; Chen, J.; Xu, N.; Gao, H. *J. Mater. Chem.* **2010**, *20*, 2197–2205.

234. Cao, L.; Zhang, Z.; Sun, L.; Gao, G.; He, M.; Wang, Y.; Li, Y.; Zhang, X.; Li, G.; Zhang, J.; Wang, W. *Adv. Mater.* **2001**, *13*, 1701–1704.

235. Chen, X.; Wu, P.; Rousseas, M.; Okawa, D.; Gartner, Z.; Zettl, A.; Bertozzi, C. R. *J. Am. Chem. Soc.* **2009**, *131*, 890–891.

236. Ciofani, G.; Raffa, V.; Menciassi, A.; Cuschieri, A. *Biotech. Bioeng.* **2008**, *101*, 850–858.

237. Ciofani, G.; Raffa, V.; Menciassi, A. *Nanoscale Res. Lett.* **2009**, *4*, 113–121.

238. Alexiou, C.; Arnold, W.; Klein, R. J.; Parak, F. G.; Hulin, P.; Bergemann, C.; Erhart, W.; Wagenpfell, S.; Lübbe, A. S. *Cancer Res.* **2000**, *60*, 6641–6648.

239. Lübbe, A. S.; Bergemann, C.; Riess, H.; Schriever, F.; Reichardt, P.; Possinger, K.; Matthias, M.; Dorken, B.; Herrmann, F.; Gurtler, R.; Hohenberger, P.; Haas, N.; Sohr, R.; Sander, B.; Lemke, A.-J.; Ohlendorf, D.; Huhnt, W.; Huhn, D. *Cancer Res.* **1996**, *56*, 4686–4693.

240. Pulfer, S. K.; Ciccotto, S. L.; Gallo, J. M. *J. Neuro-Oncol.* **1999**, *41*, 99.

241. Jurgons, R.; Seliger, C.; Hilpert, A.; Trahms, L.; Odenbach, S.; Alexiou, C. *J. Phys.: Condens. Matter* **2006**, *18*, S2893–S2902.

242. (a) Ciofani, G.; Raffa, V.; Yu, J.; Chen, Y.; Obata, Y.; Takeoka, S.; Menciassi, A.; Cuschiere, A. *Curr. Nanosci.* **2009**, *5*, 33–38. (b) Cofani G.; Raffa, V.; Meniassi, A.; Cuschieri, A. *Nanoscale Res. Lett.* **2009**, *4*, 113–121. (c) Huang, Y.; Lin, J.; Bando, Y.; Tang, C.; Zhi, C.; Shi, Y.; Takayama-Muromachi, E.; Golberg, D., *J. Mater. Chem.* **2010**, DOI: 10.1039/b916971g.

243. Zhu, Y.; Yongxiang, L.; Zhun, Z. Y.; Jia, L.; Maguire, J. A.; Hosmane, N. S. *J. Nanomater.* **2010** (*ID 409320*), p. 8.

244. Yang, W.; Barth, R. F.; Rotaru, J. H.; Moeschberger, M. L.; Joel, D. D.; Nawrocky, M. M.; Goodman, J. H.; Soloway, A. H. *Int. J. Radiat. Oncol. Biol. Phys.* **1997**, *37*(3), 663–672.

245. Khokhlov, V. F.; Yashkin, P. N.; Silin, D. I.; Djorova, E. S.; Lawaczeck, R. *Acad. Radiol.* **1995**, *2*(5), 392–398.

246. Crossley, E. L.; Aitken, J. B.; Vogt, S.; Harris, H. H.; Rendina, L. M. *Angew. Chem., Int. Ed. Engl.* **2010**, *49*, 1231–1233.

247. Faure, A. C.; Dufort, S.; Josserand, V.; Perriat, P.; Coll, J. L.; Roux, S.; Tillement, O. *Small.* **2009**, *5*(22), 2565–2575.

248. Arrais, A.; Botta, M.; Avedano, S.; Giovenzana, G. B.; Gianolio, E.; Boccaleri, E.; Stanghellini, P. L.; Aime, S. *Chem. Commun.* **2008**, 5936–5938.

249. Sitharaman, B.; Kissell, K. R.; Hartman, K. B.; Tran, L. A.; Baikalov, A.; Rusakova, I.; Sun, Y.; Khant, H. A.; Ludtke, S. J.; Chiu, W.; Laus, S.; Tóth, É.; Helm, L.; Merbach, A. E.; Wilson, L. J. *Chem. Commun.* **2005**, 3915–3917.

250. Paunesku, T.; Ke, T.; Dharmakumar. R.; Mascheri, N.; Wu, A.; Lai, B.; Vogt, S.; Maser, J.; Thurn, K.; Szolc-Kowalska, B.; Larson, A.; Bergan, R. C.; Omary, R.; Li, D.; Lu, Z. R.; Woloschak, G. E. *Nanomedicine* **2008**, *4*(3), 201–207.

251. Matsumura, A.; Zhang, T.; Yamamoto, T.; Yoshida, F.; Sakurai, Y.; Shimojo, N.; Nose, T. *Anticancer Res.* **2003**, *23*(3B), 2451–2456.

252. Hofmann, B.; Fischer, C. O.; Lawaczeck, R.; Platzek, J.; Semmler, W. *Invest. Radiol.* **1999**, *34*(2), 126–133.

253. Fujimoto, T.; Ichikawa, H.; Akisue, T.; Fujita, I.; Kishimoto, K.; Hara, H.; Imabori, M.; Kawamitsu, H.; Sharma, P.; Brown, S. C.; Moudgil, B. M.; Fujii, M.; Yamamoto, T.; Kurosaka, M.; Fukumori, Y. *Appl. Radiat. Isot.* **2009**, *67(7–8 Suppl)*, S355–S358.

254. Saha, T. K.; Ichikawa, H.; Fukumori, Y. *Carbohydr. Res.* **2006**, *341*(17), 2835–2841.

255. Le, U. M.; Shaker, D. S.; Sloat, B. R.; Cui, Z. *Drug. Dev. Ind. Pharm.* **2008**, *34*(4), 413–418.

256. Le, U. M.; Cui, Z. *Int. J. Pharm.* **2006**, *312*(1–2), 105–112.

257. Ichikawa, H.; Watanabe, T.; Tokumitsu, H.; Fukumori, Y. *Curr. Drug Deliv.* **2007**, *4*(2), 131–140.

258. Watanabe, T.; Ichikawa, H.; Fukumori, Y. *Eur. J. Pharm. Biopharm.* **2002**, *54*(2), 119–124.

259. Dierling, A. M.; Sloat, B. R.; Cui, Z. *Eur. J. Pharm. Biopharm.* **2006**, *62*(3), 275–281.

260. Oyewumi, M. O.; Mumper, R. J. *Int. J. Pharm.* **2003**, *251*(1–2), 85–97.

261. Oyewumi, M. O.; Yokel, R. A.; Jay, M.; Coakley, T.; Mumper, R.J. *J. Contr. Release.* **2004**, *95*(3), 613–626.

262. Kato, Y.; Onishi, H.; Machida, Y. *Curr. Pharm. Biotechnol.* **2003**, *4*(5), 303–309.

263. Oyewumi, M. O.; Liu, S.; Moscow, J. A.; Mumper, R. J. *Bioconjugate Chem.* **2003**, *14*, 404–411.

264. Shahbazi-Gahrouei, D.; Williams, M.; Rizvi, S.; Allen, B.J. *J. Magn. Reson. Imaging* **2001**, *14*(2), 169–174.

265. Kobayashi, H.; Kawamoto, S.; Saga, T.; Sato, N.; Ishimori, T.; Konishi, J.; Ono, K.; Togashi, K.; Brechbiel, M. W. *Bioconjugate Chem.* **2001**, *12*, 587–593.

266. Aime, S.; Barge, A.; Crivello, A.; Deagostino, A.; Gobetto, R.; Nervi, C.; Prandi, C.; Toppino, A.; Venturello, P. *Org. Biomol. Chem.* **2008**, *6*, 4460–4466.

267. Matsumura, A.; Zhang, T.; Nakai, K.; Endo, K.; Kumada, H.; Yamamoto, T.; Yoshida, F.; Sakurai, Y.; Tamamoto, K.; Nose, T. *J. Exp. Clin. Cancer Res.* **2005**, *24*, 93–98.

268. Harling, O. K.; Riley, K. J. *J. Neuro-Oncol.* **2003**, *62*, 7–17.

269. Riley, K. J.; Binns, P. J.; Harling, O. K. *Phys. Med. Biol.* **2004**, *49*, 3725–3735.

270. Binns, P. J.; Riley, K. J.; Harling, O. K. *Radiat. Res.* **2005**, *164*, 212–220.

271. Taken from http://web.mit.edu/nrl/www/bnct/1nfo/description/description.html.

272. Bisceglie, E.; Colangelo, P.; Colonna, N.; Santorelli, P.; Variale, V. *Phys. Med. Biol.* **2000**, *45*, 49–58.

273. (a) Hendricks, J. S.; Adams, K. J.; Booth, T. E.; Briesmeister, J. F.; Carter, L. L.; Cox, L. J.; Favorite, J. A.; Forster, R. A.; McKinney, G. W.; Prael, R. E. *Appl. Radiat. Isot.* **2000**, *53*, 857–861. (b) A General Monte Carlo N-particle Transport Code, version 5, LA-UR-03-1987, Los Alamos National Laboratory, April 2003, http://mcnp-green.lanl.gov/.

274. (a) Nigg, D. W. *J. Neuro-Oncol.* **2003**, *62*, 75–86. (b) Wessol, D. E.; Wemple, C. A.; Wheeler, F. J.; Harkin, G. J.; Frandsen, M. W.; Albright, C. L.; Cohen, M. T.; Rossmeier, M. B.; Cogliati, J. J. *SERA: Simulation Environment for Radiotherapy Applications User's Manual Version 1CO*, 2002, http://www.inl.gov/technicalpublications/Documents/3157074.pdf.

275. (a) Gonzalez, S. J.; Sants Cruz, G. A.; Kiger, W. S. III; Palmer, M. R.; Busse, P. M.; Zamenhof, R. G. In *Research and Development in Neutron Capture Therapy*; Sauerwein, W.; Moss, R. L.; Wittig, A., Eds.; Monduzzi Editore, Bologna, Italy, **2002**, pp. 557–561. (b) Kiger, W. S. III; Zamenhof, R. G.; Solares, G. R.; Redmond, E. L. II; Yam, C.-S. *Trans. Am. Nucl. Soc.* **1996**, *75*, 38–39.

276. (a) Li, H. S.; Liu, Y.-W. H.; Lee, C. Y.; Lin, T. Y.; Hsu, F. Y. *Appl. Radiat. Isot.* **2009**, *67*, 5122–5125. (b) Deng, L.; Li, G.; Ye, T.; Chen, C.; Zhang, B.; Zhang, L. *J. Nucl. Sci. Tech.* **2007**, *44*, 1518–1525.

277. Raaijmakers, C. P. J.; Nottelman, E. L.; Mijnheer, B. J. *Phys. Med. Biol.* **2000**, *45*, 2353–2361.

278. Gambarini, G.; Colli, V.; Gay, S.; Petrovich, C.; Rosi, G.; Valente, M. 5th International Conference on Isotopes, Brussels, Belgium, Apr 25–29, 2005.

279. Binns, P. J.; Riley, K. J.; Harling, O. K. *Radiat. Res.* **2005**, *164*, 212–220.

280. Sepälä, T.; Vähätalo, J.; Auterinen, I.; Kosunen, A.; Nigg, D. W.; Wheeler, F. J.; Savolainen, S. *Radiat. Phys. Chem.* **1999**, *55*, 239–246.

281. Harling, O. K. *Appl. Radiat. Isot.* **2009**, *67*, S7–S11.

282. Blue, T. E.; Yanch, J. C. *J. Neuro-Oncol.* **2003**, *62*, 19–31.

283. Zimin, S.; Allen, B. J. *Phys. Med. Biol.* **2000**, *45*, 59–67.

284. Tanaka, H.; Sakurai, Y.; Suzuki, M.; Takata, T.; Masunaga, S.; Kinashi, Y.; Kashino, G.; Liu, Y.; Mitsumoto, T.; Yajima, S.; Tsutsui, H.; Takada, M.; Maruuhashi, A.; Ono, K. *Appl. Radiat. Isot.* **2009**, *67*, S258–S261.

285. Suzuki, M.; Tanaka, H.; Sakurai, Y.; Kashino, G.; Yong, L.; Masunaga, S.; Kinashi, Y.; Mitsumoto, T.; Yajima, S.; Tsutsui, H.; Sato, T.; Maruhashi, A. *Radiat. Oncol.* **2009**, *92*, 89–95.

286. Yanch, J. C.; Zhou, X.-L.; Shefer, R. E.; Klinkowstein, R. E. *Med. Phys.* **1992**, *19*, 709–721.

287. Kreiner, A. J.; Di Paolo, H.; Burlon, A. A.; Kesque, J. M.; Valda, A. A.; Debray, M. E.; Giboudot, Y.; Levinas, P.; Fraiman, M.; Romeo, V.; Somacal, H. R.; Minsky, D. M. In VII Latin American Symposium on Nuclear Physics and Applications; Alarcon, R.; Cole, P. L.; Djalali, C.; Umeres, F. Eds.; American Institute of Physics, 2007, pp. 17–24.

288. Bayanov, B.; Belov, V.; Taskaev, S. *J. Phys.: Conf. Ser.* **2006**, 141, 460–465.

289. Hirshfeld, H.; Silverman, I.; Arenshtam. A.; Kifel. D.; Nagler, A. *Nucl. Instr. Meth. Phys. Res. A* **2006**, *562*, 903–905.

290. Willis, C.; Lenz, J.; Swenson, D. *Proc. LINAC08* **2008**, 223–225.

291. Kuznetsov, A. S.; Malyshin, G. N.; Makarov, A. N.; Sorokin, I. N.; Sulyaev, Y. S.; Taskaev, Y. *Tech. Phys. Lett.* **2009**, *35*, 346–348.

292. Colona, N.; Beaulieu, L.; Phair, G.; Wozniak, G. J.; Moretto, L. G.; Chu, W. T.; Ludewigt, B. A. *Med. Phys.* **1999**, *26*, 793–798.

293. Verbeke, J. M.; Lee, Y.; Leung, K. N.; Vujic, J.; Williams, M. D.; Wu, L. K.; Zahir, N. *Proc. Part. Accel. Conf.* **1999**, 2540–2542.

294. Reijonen, J.; Gicquel, F.; Hahto, S. K.; King, M.; Lou, T.-P.; Leung, K.-N.; *Appl. Radiat. Isot.* **2005**, *63*, 757–763.

295. Lou, T. P.; Vujic, J. L.; Koivunoro, H.; Reijonen, J.; Leung, K. N. *Fission Multipliers for D-D/D-T Neutron Generators, Paper LBNL-52417*, 2003, can be downloaded at http://repositories.cdlib.org/lbnl/LBNL-52417.

296. Yonai, S.; Baba, M.; Nakamura, T.; Yokobori, H.; Tahara, Y. *J. Nucl. Sci. Tech.* **2008**, *45*, 378–383.

297. See: *Califorium-252: Supply and Funding Status* ; Nuclear Energy Institute 2009, can be downloaded at http:// www.nei.org/.../**californium**-**252**-supply-and-funding-status/.

298. Ghassoun, J.; Chkillou, B.; Jehouani, A. *Appl. Radiat. Isot.* **2009**, *67*, 560–564.

299. Yanch, J. C.; Zamenhof, R. G. *J. Radiat. Res.* **1992**, *131*, 249–256.

300. Gallion, H. H.; Maruyama, Y.; Van Nagell, Jr, J. R.; Donaldson, E. S.; Rowley, K. C.; Yoneda, J.; Beach, J. L.; D. E.; Kryscio, R. J. *Cancer* **1987**, *59*, 1709–1712.

301. Burmiester, J.; Kota, C.; Maughan, R. L. *Radiat. Res.* **2005**, *164*, 312–318.

302. Stelzer, K. J.; Lindsley, K. L.; Cho, P. S.; Laramore, G. E.; Griffin, T. W. *Radiat. Proct. Dosim.* **1997**, *70*, 471–475.

303. Rogus, R. D.; Harling, O. K.; Yanch, J. C. *Med. Phys.* **1994**, *21*, 1611–1625.

304. Caswell, R. S.; Coyne, J. J.; Randolph, M. L. *Radiat. Res.* **1980**, *83*, 217–254.

305. Coderre, J. A.; Makar, M. S.; Micca, P. L.; Nawrocky, M. M.; Liu, H. B.; Joel, D. D.; Slatkin, D. N.; Amols, H. I. *J. Radiat. Ocol. Biol. Phys.* **1993**, *27*, 1121–1129.

306. Morris, G. M.; Coderre, J. A.; Hopwell, J. W.; Micca, P. L.; Rezvani, M. *Radiother. Oncol.* **1994**, *32*, 144–153.

307. Morris, G. M.; Coderre, J. A.; Hopwell, J. W.; Micca, P. L.; Nawrocky, M. M.; Liu, H. B.; Bywaters, A. *Radiother. Oncol.* **1994**, *32*, 249–255.

308. Morris, G. M.; Smith, D. R.; Patel, H.; Chandra, S.; Morrison, G. H.; Hopwell, J. W.; Rezvani, M.; Micca, P. L.; Coderre, J. A. *Br. J. Cancer* **2000**, *82*, 1764–1771.

309. Schreiner, L. J. *J. Phys.: Conf. Ser.* **2004**, *3*, 9–21.

310. Chaterjee, A.; Magee, J. L. *J. Phys. Chem.* **1980**, *84*, 3537–3543.

311. Vanossi, E.; Carrara, M.; Gambarini, G.; Negri, A.; Mariani, M. *Appl. Radiat. Isot.* **2009**, *67*, 5195–5198.

312. Raaijmakers, C. P. J.; Konijnenberg, M. W.; Verhagen, H. W.; Mijnheer, B. J. *Med. Phys.* **1995**, *22*, 321–329.

313. Rhoades, W. A. *ORNL Report ORNL/TM-13221*, Oak Ridge, TN, 1996.

314. Barth, R. F.; Grecula, J. C.; Yang, W.; Rotaru, J. H.; Nawrocky, M.; Gupta, N.; Albertson, B. J.; Ferketich, A.; Moeschberger, M. L.; Coderre, J. A.; Rofstrad, E. K. *Int. J. Radiat. Oncol. Biol. Phys.* **2004**, *58*, 267–277.

315. Wemple, C. A.; Wessol, D. E.; Nigg, D. W.; Cogliati, J. J.; Milvich, M. L.; Fredrickson, C.; Perkins, M.; Harkin, G. J. *Appl. Radiat. Isot.* **2004**, *61*, 745–752.

316. Takagaki, M.; Powell, W.; Sood, A.; Spielvogel, B. F.; Hosmane, N. S.; Kirihata, M.; Ono, K.; Masunaga, S. I.; Kinashi, Y.; Miyatake, S. I.; Hashimoto, N. *Radiat. Res.* **2001**, *156*(1), 118–122.

317. Mitin, M.; Kulakov, V. N.; Khokhlov, V. F.; Sheino, I. N.; Arnopolskaya, A. M.; Kozlovskaya, N. G.; Zaitsev, K. N.; Portnov, A. A. *Appl. Radiat. Isot.* **2009**, *67*, FS299–FS301.

318. Cerullo, N.; Bufalino, D.; Daquino, G. *Appl. Radiat. Isot.* **2009**, *67*, S157–S160.

319. Kullberg, E. B.; Carlsson, J.; Edwards, K.; Capala, J.; Sjöberg, S.; Gedda, L. *Int. J. Oncol.* **2003**, *23*, 461–467.

320. Pan, X.; Wu, G.; Yang, W.; Barth, R. F.; Tjarks, W.; Lee, R. J. *Bioconjugate Chem.* **2007**, *18*, 101–108.

321. Schnyder, A.; Huwyler, J. *NeuroRx* **2005**, *2*, 99–107.

322. Zhu, Y.; Maguire, J. A.; Hosmane, N. S. In *Boron Science: New Technologies & Applications*; Hosmane, N. S., Ed.; CRC Press: Boca Raton, FL, 2011, Chapt. 4.2.

323. Nakamura, H. In *Boron Science: New Technologies & Applications*; Hosmane, N. S., Ed.; CRC Press: Boca Raton, FL, 2011, Chapt. 4.1.

324. (a) Renner, M. W.; Miura, M.; Easson, M. W.; Vicente, M. G. H. *Curr. Med. Chem., AntiCancer Agents* **2006**, *6*, 145–157. (b) Sibrian-Vazquez, M.;

Vicente, M. G. H. In *Handbook of Porphyrin Science*; Kadish, K. M.; Smith, K. M.; Guilard, R. Eds.; World Scientific Publishers, Singapore, 2010, Vol. 4, Chap. 18, pp. 191–248.

325. Sibrian-Vazquez, M.; Vicente, M. G. H. In *Boron Science: New Technologies & Applications*; Hosmane, N. S., Ed.; CRC Press: Boca Raton, FL, 2011, Chap. 4.4.

326. Rosenthal, M. A.; Kavar, B.; Hill, J. S.; Morgan, D. J.; Nation, R. L.; Stylli, S. S.; Basser, R. L.; Uren, S.; Geldard, H.; Green. M. D.; Kahl, S. B.; Kaye, A. H. *J. Clin. Oncol.* **2001**, *19*, 519–524.

327. Barth, R. F.; Yang, W.; Al-Madhoun, A. S.; Johnsamuel, J.; Byun, Y.; Chandra, S.; Smith, D. R.; Tkarks, W.; Eriksson, S. *Cancer Res.* **2004**, *64*, 6287–6295.

328. Byun, Y.; Thirumamagal, B. T. S.; Yang, W.; Eriksson, S.; Barth, R. F.; Tjarks, W. *J. Med. Chem.* **2006**, *49*, 5513–5523.

329. Byun, Y.; Yan, J.; Al-Madhoun, A. S.; Johnsamuel, J.; Yang, W.; Barth, R. F.; Erksson, S.; Tjarks, W. *J. Med. Chem.* **2005**, *48*, 1188–1198.

330. Martin, R. F.; D'Cunha, G.; Pardee, M.; Allen, B. J. *Pigment Cell Res.* **1989**, *2*(4), 330–332.

331. (a) Akine, Y.; Tokita, N.; Matsumoto, T.; Oyama, H.; Egawa, S.; Aizawa, O. *Strahlenther Onkol.* **1990**, *166*(12), 831–833. (b) Akine, Y.; Tokita, N.; Tokuuye, K.; Satoh, M.; Fukumori, Y.; Tokumitsu, H.; Kanamori, R.; Kobayashi, T.; Kanda, K. *J. Cancer Res. Clin. Oncol.* **1992**, *119*(2), 71–73.

332. Oda, Y.; Takagaki, M.; Miyatake, S.; Kikuchi, H. In *Intraoperative Radiation Therapy*; Abe, M.; Takagashi, M., Eds.; Pergamon Press: Oxford, 1988, p. 156.

333. De Stasio, G.; Casalbore, P.; Pallini, R.; Gilbert, B.; Sanità, F.; Ciotti, M. T.; Rosi, G.; Festinesi, A.; Larocca, L. M.; Rinelli, A.; Perret, D.; Mogk, D. W.; Perfetti, P.; Mehta, M. P.; Mercanti, D. *Cancer Res.* **2001**, *61*(10), 4272–4277.

334. Franken, N. A.; Bergs, J. W.; Kok, T. T.; Kuperus, R. R.; Stecher-Rasmussen, F.; Haveman, J.; Van Bree, C.; Stalpers, L. J. *Oncol. Rep.* **2006**, *15*(3), 715–720.

335. Masunaga, S.; Ono, K.; Sakurai, Y.; Suzuki, M.; Takagaki, M.; Kobayashi, T.; Kinashi, Y.; Akaboshi, M. *Jpn. J. Cancer Res.* **1998**, *89*(1), 81–88.

336. Yasui, L. S.; Andorf, C.; Schneider, L.; Kroc, T.; Lennox, A.; Saroja, K, R. *Int. J. Radiat. Biol.* **2008**, *84*(12), 1130–1139.

337. Akine, Y.; Tokita, N.; Tokuuye, K.; Satoh, M.; Churei, H.; Le Pechoux, C.; Kobayashi, T.; Kanda, K. *Jpn. J. Cancer Res.* **1993**, *84*(8), 841–843.

338. Le, U. M.; Shaker, D. S.; Sloat, B. R.; Cui, Z. *Drug Dev. Ind. Pharm.* **2008,** *34*(4), 413–418.

339. Mitin, V. N.; Kulakov, V. N.; Khokhlov, V. F.; Sheino, I. N.; Arnopolskaya, A. M.; Kozlovskaya, N. G.; Zaitsev, K. N.; Portnov, A. A. *Appl. Radiat. Isot.* **2009,** *67,* S299–S301.

340. Aime, S.; Barge, A.; Crivello, A.; Deagostino, A.; Gobetto, R.; Nervi, C.; Prandi, C.; Toppino, A.; Venturello, P. *Org. Biomol. Chem.* **2008,** *6*(23), 4460–4466.

341. Takahashi, K.; Nakamura, H.; Furumoto, S.; Yamamoto, K.; Fukuda, H.; Matsumura, A.; Yamamoto, Y. *Bioorg. Med. Chem.* **2005,** *13*(3), 735–743.

342. Culbertson, C. N.; Jevremovic, T. *Phys. Med. Biol.* **2003,** *48*(23), 3943–3959.

343. Nakagawa, Y.; Pooh, K.; Kobayashi, T.; Kageji, T.; Uyama, S.; Matsumura, A.; Kumada, H. *Int. J. Neuro-Oncol.* **2003,** *62,* 87–99.

344. Laramore, G. E.; Wooton, P.; Livesey, J. C.; Wilbur, D. S.; Risler, R.; Phillips, M.; Jacky, J.; Buchholz, T. A.; Griffin, T. W.; Brossard, S. *Int. J. Radiat. Oncol. Biol. Phys.* **1994,** *28,* 1135–1142.

345. Wittig, A.; Hidgehty, K.; Paquis, P.; Heimans, J.; Vos, M.; Goetz, C.; Haselberger, K.; Grouchulla, F.; Moss, R.; Morisssey, J.; Bourhis-Martin, E.; Rassow, J.; Stecher-Rasmussen, F.; Turowski, B.; Wiestler, O. D.; DeVries, M.; Frankhauser, H.; Gabel, D.; Sauerwein, W. In *Research and Development in Neutron Capture Therapy*; Sauerwein, W.; Moss, R.; Wittig, A. Eds.; Monduzzi Editore: Bologna, **2002,** pp. 82–11931117–1112.

346. Diaz, A. Z. *J. Neuro-Oncol.* **2003,** *62,* 101–109.

347. Coderre, J. A.; Hopewell, J. W.; Turcotte, J. C.; Riley, K. J.; Binns, P. J.; Kiger III, S.; Harling, O. K. *Appl. Radiat. Isot.* **2004,** *61,* 1083–1087.

348. Chanana, A. D.; Capala, J.; Chadha, M.; Coderre, J. A.; Diaz, A. Z.; Elowitz, E. H.; Iwai, J.; Joel, D. D.; Liu, H. B.; Ma, R. *Neurosurgery* **1999,** *44,* 1182–1193.

349. Busse, P. M.; Harling, O. K.; Palmer, M. R.; Kiger III, W. S.; Kaplan, J.; Kaplan, I.; Chuang, C. F.; Goorley, J. T.; Riley, K. J.; Newton, T. H.; Santa Cruz, G. A.; Lu, X.-Q.; Zamenhof, R. G. *J. Neuro-Oncol.* **2003,** *62,* 111–121.

350. Capala, J.; Stenstam, B. H.; Sköld, K.; Rosenschöld, P. M.; Giusti, V.; Persson, C.; Wallin, E.; Brun, A.; Franzen, L.; Carlsson, J.; Salford, L.; Ceberg, C.; Persson, B.; Pellettieri, L.; Henriksson, R. *J. Neuro-Oncol.* **2003,** *62,* 135–144.

351. Joensuu, H.; Kankaanranta, L.; Seppälä, T.; Auterinen, I.; Kallio, M.; Kulvik, M.; Laasko, J.; Vähätalo, J.; Kortesniemi, M.; Kotiluoto, P.; Serén, T.; Karila, J.; Brander, A.; Järviluoma, E.; Ryynänen, P.; Paetau, A.; Roujonen, I.; Minn, H.; Tenhunen, M.; Jääskeläinen, J.; Färkkilä, M.; Savolainen, S. *J. Neuro-Oncol.* **2003**, *62*, 123–134.

352. Burian, J.; Marek, M.; Rataj, J.; *et al.* In *Research and Development in Neutron Capture Therapy*; Sauerwein, W.; Moss, R.; Wittig, A. Eds.; Monduzzi Editore: Bologna, **2002**, pp. 1107–1112.

353. Kawabata, S.; Miyatake, S.-I.; Nonoguchi, N.; Iida, K.; Miyata, S.; Yokoyama, K.; Doi, A.; Kurdo, Y.; Kuroiwa, T.; Michiue, H.; Kumada, H.; Kirihata, M.; Imahori, Y.; Maruhashi, A.; Sakurai, Y.; Suzuki, M.; Masunaga, S.-I.; Ono, K. Proceedings of the 13th International Congress on Neutron Capture Therapy: A New Option Against Cancer; Zonta, A.; Altierii, S.; Roveda, L.; Barth, R., Eds; ENEA, 2008, pp. 13–16.

354. Nakagawa, Y.; Kageji, T.; Mizobuchi, Y.; Nagahiro, S.; Kumada, H. Proceedings of the 13th International Congress on Neutron Capture Therapy: A New Option Against Cancer; Zonta, A.; Altierii, S.; Roveda, L.; Barth, R., Eds.; ENEA, 2008, pp. 17–20.

355. Laramore, G. E.; Spence, A. M. *Int. J. Radiat. Oncol. Biol. Phys.* **1996**, *36*, 241–246.

356. Matsuda, M.; Yamamoto, T.; Kumada, H.; Nakai, K.; Shirakawa, M.; Tsurubuchi, T.; Matsumura, A. *Appl. Radiat. Isot.* **2009**, *67*, S19–S21.

357. Yamamoto, T.; Tsuboi, K. *Brain Nerve.* **2009**, *61*, 855–866.

358. Kawabata, S.; Miyatake, S.; Kuroiwa, T.; Yokoyama, K.; Doi, A.; Iida, K.; Miyata, S.; Nonoguchi, N.; Michiue, H.; Takahashi, M.; Inomata, T.; Imahori, Y.; Kirihata, M.; Sakurai, Y.; Maruhashi, A.; Kumada, H.; Ono, K. *J. Radiat. Res. (Tokyo)* **2009**, *50*, 51–60.

359. Kato, I.; Ono, K.; Sakuria, Y.; Ohmae. M.; Maruhashi, A.; Imahori, Y.; Kirihaa, M.; Nakazawa, M.; Yura, Y. *Appl. Radiat. Isot.* **2004**, *61*, 1069–1073.

360. Kato, I.; Fujita, Y.; Maruhashi, A.; Kumada, H.; Ohmae, M.; Kirihata, M.; Imahori, Y.; Suzuki, M.; Sakrai, Y.; Sumi, T.; Iwai, S.; Nakazawa, M.; Murata, I.; Miyamaru, H.; Ono, K. *Appl. Radiat. Isot.* **2009**, *67*, S37–S42.

361. The authors express thanks to Dr Itsuro Kato, Osaka University for permission to show Figs. 8.5 and 8.6. [Correspondence: Dr Itsuro Kato, MD, PhD, Department of Oral and Maxillofacial Surgery II, Graduate School of Dentistry, Osaka University, 1–8 Yamada-oka, Suita, Osaka 565–0871, Japan (katoitsu@dent.osaka-u.ac.jp)].

362. Joensun, H.; Knnkaanrante, L.; Seppälä, T.; Auterinen, I.; Kallio, M.; Kulvik, M.; Laakso, J.; Vähätalo, J.; Kortesniemi, M.; Kotiluoto, P.; Serén, T.; Karila, J.; Brander, A.; Järviluoma, E.; Ryynänen, P.; Paetau, A.; RuoKonen, I.; Minn, H.; Tenhunen, M.; Jääskeläinen, J.; Farkkilä, M.; Savolainen, S. *J. Neuro-Oncol.* **2003**, *62*, 123–134.

363. Kankaanranta, L.; Seppälä, T.; Koivunoro, H.; Saarilahti, K.; Atula, T.; Collan, J.; Salli, E.; Kortesniemi, M.; Uusi-Simola, J.; Mäkitie, A.; Seppänen, M.; Minn, H.; Kotiluoto, P.; Auterinen, I.; Savolainen, S.; Kouri, M.; Joensuu, H. *Int. J. Radiat. Oncol. Biol. Phys.* **2007**, *69*, 475–482.

364. Capala, J.; Stenstam, B. H.; Sköld, K.; Rosenschölf, P. M.; Giusti, V.; Persson, C.; Wallin, E.; Brun, A.; Franzen, L.; Carlsson, J.; Salford, L.; Ceberg, C.; Persson, B.; Pellettieri, L.; Henriksson, R. *J. Neuro-Oncol.* **2003**, *62*, 135–144.

365. Henriksson, R.; Capala, J.; Michanek, A.; Lindahl, S.-A.; Salford, L. G.; Franzén, L.; Blomquist, E.; Westlin, J.-E.; Bergenheim, A. T. *Radiother. Oncol.* **2008**, *88*, 183–191.

366. Menéndez, P. R.; Roth, B. M. C.; Pereira, M. D.; Casal, M. R.; González, S. J.; Feld, D. B.; Santa-Cruz, G. A.; Kessler, J.; Longhino, J.; Blaumann, H.; Rebagliati, R. J.; Larrieu, O. A. G.; Fernández, C.; Nievas, S. I.; Liberman, S. J. *Appl. Radiat. Isot.* **2009**, *67*, 550–553.

367. Zonta, A.; Prati, U.; Roveda, L.; Ferrari, C.; Zonta, S.; Clerici, A. M.; Zonta, C.; Pinelli, T.; Fossati, F.; Altieri, S.; Bortolussi, S.; Bruschi, P.; Nano, R.; Barni, S.; Chiara, P.; Mazzini, G. *J. Phys. Conf. Ser.* **2006**, *41*, 484–495.

368. Wittig, A.; Malago, M.; Collette, L.; Huiskamp, R.; Bührmann, S.; Nievaat, V.; Kaiser, G. M.; Jöckel, K.-H.; Schmid, K. W.; Ortmann, U.; Sauerwein, W. A. *Int. J. Cancer* **2008**, *122*, 1164–1171.

369. Suzuki, M.; Sakurai, Y.; Hagiwara, S.; Masunaga, S.; Kinashi, Y.; Nagata, K.; Maruhashi, A.; Kudo, M.; Ono, K. *Jpn. J. Clin. Oncol.* **2007**, *37*, 376–381.

370. Suzuki, M.; Endo, K.; Satoh, H.; Sakura, Y.; Kumada, H.; Kimura, H.; Masunaga, S.; Kinashi, Y.; Nagata, K.; Maruhashi, A.; Ono, K. *Radiother. Oncol.* **2008**, *88*, 192–195.

371. Tokumitsu, H.; Hiratsuka, J.; Sakurai, Y.; Kobayashi, T.; Ichikawa, H.; Fukumori, Y. *Cancer Lett.* **2000**, *150*, 177–182.

372. (a) Butt, M.; Connolly, D.; Lip, G. Y. *Future Cardiol.* **2009**, *5*, 141. (b) Kiernan, T. J.; Yan, B. P.; Cruz-Gonzalez, I.; Cubeddu, R. J.; Caldera, A.; Kiernan, G. D.; Gupta, V. *Cardiovasc. Hematol. Agents Med. Chem.* **2008**, *6*, 116. (c) Zargham, R. *Clin Sci (Lond)* **2008**, *114*, 257.

373. King III, S. B.; Williams, D. O.; Chougule, P.; Klein, J. L.; Waksman, R.; Hilstead, R.; Macdonald, J.; Anderberg, K.; Crocker, J. R. *Circulation* **1998**, *97*, 2025.

374. Sadek, M.; Cayne, N. S.; Shin, H. J.; Turnbull, I. C.; Martin, M. L.; Faries, P. J. *J. Vasc. Surg.* **2009**, *50*, 1308.

375. Takagaki, M.; Hosmane, N. S. *AINO J.* **2007**, *6*, 39–44.

376. Enger, S. A.; Rezaei, A.; Munck af Rosenschöld, P.; Lundqvist, H. *Med. Phys.* **2006**, *33*, 46–51.

377. Shih, J. L.; Brugger, R. M. *Med. Phys.* **1992**, *19*, 369–375.

378. De Stasio, G.; Rajesh, D.; Casalbore, P.; Daniels, M. J.; Erhardt, R. J.; Frazer, B. H.; Wiese, L. M.; Richter, K. L.; Sonderegger, B. R.; Gilbert, B.; Schaub, S.; Cannara, R. J.; Crawford, J. F.; Gilles, M. K.; Tyliszczak, T.; Fowler, J. F.; Larocca, L. M.; Howard, S. P.; Mercanti, D.; Mehta, M. P.; Pallini, R. *Neuro. Res.* **2005**, *27*, 387–398.

379. Barth, R. F. *J. Neuro. Oncol.* **2003**, *62*, 1–5.

380. Barth, R. F. *Appl. Radiat. Isot.* **2009**, *67*, S3–S6.

381. Barth, R. F.; Corderre, J. A.; Vicente, M. G. H.; Blue, T. E. *Clin. Cancer Res.* **2005**, *11*, 3987–4002.

382. Hawthorne, M. F.; Lee, M. W. *J. Neuro. Oncol.* **2003**, *62*, 33–45.

383. Soloway, A. H.; Tjarks, W.; Barnum, B. A.; Rong, F.-G.; Barth, R. F.; Codogni, I. M.; Wilson, J. G. *Chem. Rev.* **1998**, *98*, 1515–1562.

384. Linz, U. *Technol. Cancer Res. Treat.* **2008**, *7*, 83–88.

385. Paganetti, H. *Technol. Cancer Res. Treat.* **2003**, *2*, 353–354.

386. (a) Zhu, Y.; Koh, C.-Y.; Maguire, J. A.; Hosmane, N. S. *Curr. Chem. Biol.* **2007**, *1*, 141–149. (b) Hosmane, N. S.; Zhu, Y.; Maguire, J. A.; Takagaki, M. *J. Organomet. Chem.* **2009**, *694*, 1690–1697. (c) Hosmane, N. S.; Zhu, Y.; Maguire, J. A.; Hosmane, S. N.; Chakrabarti, A. *Main Group Chem.* **2010**, *9*, 153–166.

Index